The Dragonflies of Essex

Ted Benton

John Dobson

The Dragonflies of Essex

Published by The Essex Field Club in association with Lopinga Books

With financial support from the Corporation of the City of London, Environment Agency, English Nature, Essex County Council & Defra, Essex Field Club, Essex & Suffolk Water and the Forestry Commission

First published 2007

ISBN-10: 0-905637-18-6
ISBN-13: 978-0-9056-3718-1

British Library Cataloguing-in-Publication Data
A catalogue record for this book is available from the British Library

Printed by Healeys Printers Ltd, Unit 10, The Sterling Complex, Farthing Road, Ipswich IP1 5AP

The *Nature of Essex* Series No. 6

Contents

Foreword   vii

Acknowledgements   ix

Chapter 1   Introduction   1

Chapter 2   Where to see dragonflies in Essex   31

Chapter 3   The species   59

Chapter 4   A history of dragonfly recording in Essex   175

Appendix 1   Species formerly recorded in Essex but not recorded during the present survey   199

Appendix 2   Possible future visitors and colonists: species to look out for   209

Appendix 3   The discovery of the Small Red-eyed Damselfly in Essex   211

Appendix 4   Lesser Emperor Dragonfly   213

Appendix 5   Scientific names of plants mentioned in the text   215

Bibliography   217

Index   227

Foreword

In the last 25 years there have been considerable strides forward in our knowledge of the distribution and status of dragonflies in Britain, and in our understanding of their behaviour and ecology. To a large extent this is due to the increase in the number of people who are interested enough in dragonflies to record their presence at local sites and publish observations of their life styles. Instrumental in encouraging this increase in awareness of dragonflies has been the publication of field guides and county-based accounts of dragonflies, which have burgeoned since 1980.

One of the pioneering county guides was Ted Benton's *Dragonflies of Essex* published in 1988. This book set a bench mark in thoroughness, insightfulness and readability. Now, 19 years since the publication of the original *Dragonflies of Essex*, Ted Benton has teamed up with John Dobson and provided us with an updated account. With the publication of the new edition of *Dragonflies of Essex* we are reminded of the rich dragonfly fauna of that county, the wide variety of wetland habitats available, and also the long history of dragonfly recording, which has helped to put the distribution and status of the present species in context.

The publication of this book is timely. In the last 10-15 years there have been unprecedented changes in the British dragonfly fauna. Many of the species that were formerly restricted to southern counties of England and Wales are now breeding in northern England and Scotland, apparently in response to climate warming. Many of the species restricted to flowing waters, especially lowland rivers, are now more widespread and abundant, probably due to recent improvements in water quality and river management. There has also been a significant increase in the frequency and abundance of migrant species arriving here. Several species that were never recorded before 1990 are now regular arrivals and one species, *Erythromma viridulum*, has recently colonised this country. This species was first recorded in Essex which under-lines the significance of this county in charting the fortunes of our dragonfly fauna.

I am delighted to see the publication of this new edition of *Dragonflies of Essex*. The first edition led the way in demonstrating how local dragonfly recorders could make an important contribution to the conservation of Odo-nata by identifying key sites that support rich assemblages or rare species, by providing baseline data against which future changes can be compared, and by promoting an awareness of dragonflies to others. I wish Ted and John success with their book and in their future efforts. I anticipate this edi-tion will raise the bar still higher in what is expected of similar publications. I look forward to a third edition in 2020, by which time I am quite certain even more people will be recording dragonflies in Essex and more changes in the fauna will be apparent.

Stephen J. Brooks
Department of Entomology
The Natural History Museum
Cromwell Road
SW7 5BD

Acknowledgements

We would like to thank Steve Brooks of the Department of Entomology at The Natural History Museum for providing a foreword for this book. Steve is a member of the Journal Advisory Panel of the British Dragonfly Society and was recently General Editor of the successful British Wildlife publication *Field Guide to the Dragonflies and Damselflies of Great Britain and Ireland*.

Stephen Dewick and Richard Gerussi, the authors of the paper that in 1999 first announced the discovery of the small red-eyed damselfly in Britain, and Mark Tunmore, the editor of *Atropos*, the journal in which the paper appeared, have permitted that article to be reproduced in this volume.

Tim Bernhard has contributed the attractive painting of a four-spotted chaser.

Andrew Middleton allowed a photograph of a keeled skimmer in Epping Forest to be included amongst the colour plates of this atlas. Adrian Kettle kindly provided images of red-veined darters, photographed at Abberton Reservoir, for inclusion amongst the species accounts.

The maps were generated using the DMAP software package created by Dr Alan Morton. Peter Harvey compiled the county boundary including the location of major watercourses and we are indebted to him for this contribution. We would also like to thank Steve Cham and Adrian Parr, of the British Dragonfly Society, for their help and support.

The task of proof-reading the manuscript was undertaken by Dr Chris Gibson and Dr David Corke. We are grateful for their suggestions, which have enhanced the final appearance of this book.

List of observers

Almost one hundred people responded to the authors' coaxing and cajoling and contributed records for this project.

Special thanks are due to Colin Jupp for his comprehensive coverage of TQ59, to Andrew Middleton for his recording in Epping Forest, an area of historic significance for Essex Odonata, to Iris Cotgrove for her many records from south Essex and to Russell Neave and Simon Wood for their recording around Maldon and the Blackwater.

Many others have contributed and we apologise to anyone whose name has been inadvertently omitted.

K. Alexander	C.S. Balchin	C. Balmire	S. Banks
J.P. Bowdrey	S. Cham	D. Blakesley	D. Bridges
U.A. Broughton	P. Browne	C. Cadman	P. Charlton
J.S. Clark	I. Cotgrove	R. Cottrill	A. Cook
G. Court	J. Court	S. Cox	N. Cuming
D.J. Dagley	G. Davis	G. Ekins	A.R.G. Findlay
J. Firmin	G. Foott	L. Forsythe	R. Gardiner
K. Gash	the late C. Griffin	A. Gudgion	T.Gunton
M. Hanson	S.R. Harris	P. Harvey	I. Hawkins
S. Harris	N. Harvey	M. Heywood	P. Hodge
the late S. Hudgell	C. Huggins	M. Hunter	J.M Hurley
S. Ireland	S. Jiggins	C. Jupp	A. Kettle
A. Knowles	R.J.W. Ledgerton	D. Longe	T. Martin
A. McGeeney	A. Middleton	K. Morris	R. Neave
H. Owen	R.G. Payne	J. Pepper	S. Perry
C.W. Plant	the late G. Pyman	N.M. Rayner	A. Reynolds
P. Rhodes	J. Rowland	D. Sampson	A. Samuels
A. Sapsford	R. Seago	R. Sharp	G. Smith
P. Smith	S. Cox	R. Swain	R. Tabor
R. Tattersall	M. Telfer	A. Thomas	A. Thompson
S. & J. Torino	J.P. Tyler	D. Urquart	H. Vaughan
A. Wallington	P. Waterton	C. Watson	B. Watts
S. Wilkinson	P. & P. Wilson	G. Winn	J. Wisbey
A.C.J. Wood	D. Wood	N. Wood	S. Wood
A. Woodhouse	J. Wright	M. Wright	

Sponsors

We are grateful to those organisations that have provided financial support for this project:

The Corporation of London is the local authority for the City of London, the financial and commercial heart of Britain. However, its responsibilities extend far beyond the City boundaries and it provides a host of additional facilities for the benefit of the nation. Included amongst these is its protection of open spaces in and around London, such as Hampstead Heath, Ashtead Common, Burnham Beeches and, the largest of all, Epping Forest, straddling the borders of Essex and London. The Corporation saved Epping Forest from destruction by developments and the resulting Epping Forest Act of 1878 constituted the Corporation of London members as Conservators of Epping Forest and invested the Corporation with legal powers and duties as the owner of and managing body for the Forest. Today, the Corporation, through its Epping Forest & Open Spaces Committee, oversees the management of all the open spaces, details of which are set out in comprehensive published management plans for each area.

English Nature (now **Natural England**) is the statutory adviser to Government on nature conservation in England and promotes the conservation of England's wildlife and natural features. Its work includes selecting, establishing and managing National Nature Reserves; identifying, notifying and protecting Sites of Special Scientific Interest; providing advice and information about nature conservation, and supporting and carrying out research relevant to these functions. A major part of English Nature's work is devoted to developing and implementing Biodiversity Action Plans, at the national and local level. English Nature leads on the action plans for 59 of the 116 species identified as threatened and declining in the UK, and works to safeguard these populations and enhance the habitats that support them.

The **Environment Agency** has a wide range of duties and powers relating to environmental management and improvement in the quality of air, land and water. Part of this responsibility is to encourage the conservation of natural resources, animals and plants. As part of the government's UK Biodiversity Action Plan, the Agency is the contact point for 12 species of aquatic animal and plant including the otter and water vole. In Essex, we have undertaken collaborative surveys of these species to determine their extent in the county.

Essex County Council and **Defra** are delighted to be able to support this new volume with a grant from the Aggregates Levy Sustainability Fund. This fund introduced in 2002 aims, amongst other things, to compensate local communities for the impacts of aggregates extraction. This book will undoubtedly enable a better understanding of the habitats and food requirements of species such as the dragonfly, which is essential if new wetlands created during extraction are to be successfully formed and managed.

The **Essex Field Club** is a society for wildlife enthusiasts and people with an interest in the natural history and geology of Essex. The Club offers a programme of over 40 outdoor meetings a year, covering plants, mosses, fungi, lichens, mammals, birds, invertebrates and geology with further indoor meetings in winter. Through its Recorders, the Field Club maintains detailed scientific records for the Essex flora, fauna and geology that are important for local and national purposes and in advising on the management and protection of sites. Members receive a Newsletter three times a year, with more scientific papers published annually in the *Essex Naturalist*. The Field Club also publishers or is associated with the publication of occasional monographs.

Essex & Suffolk Water is committed to conducting the business of water supply in a sustainable manner throughout, what is, the driest region in the UK. As a responsible business, Essex & Suffolk Water aim to balance competing social, environmental and economic priorities. We do this by working in close partnership with many organisations such as local Wildlife Trusts, the Environment Agency and English Nature. These have included the development of a 100 acre reserve and the building of a visitor centre in partnership with the Essex Wildlife Trust at Hanningfield Reservoir which opened in 2000, and the successful management of key habitats at sites like Lound, Abberton Reservoir and Ormesby Broad.

Around three-quarters of England's woodlands and forests are privately owned and about one-quarter are public forests managed by the **Forestry Commission**, through its agency Forest Enterprise. England was once largely covered with woodland, but over many centuries this was cleared and used to meet the needs of an increasing population. As long as 1,000 years ago, England's woodland cover was already only 15 per cent of its land area. By the beginning of the 20th century it had reached a low point of 5 per cent. At the turn of the second millennium, woodland cover has increased but remains low at just 8 per cent. Today, woodlands and forests cover just over 1 million hectares of England, containing around 2 billion trees.

Chapter 1

Introduction

Dragonflies are among the most strikingly beautiful of all insects. Most have colourful body-patterns, while a few also have spectacularly tinted wings. Above all, their powerful and acrobatic aerial displays are an awesome sight, and trying to interpret their complex behaviour is a never-ending source of fascination. Despite widespread myths, they are completely harmless to humans (but fierce predators of other insects). Anyone who lives close to a river, stream or pond (that is, just about everyone) will have ample opportunity to spend time watching dragonflies hunting for their prey, mating, egg-laying in or around water, and, with luck, emerging from the water and beginning their lives as adult insects.

Many people will be happy just to enjoy watching, but others will have their curiosity aroused, will want to find out more – about their life-histories, their distinctive habitat requirements – and also to distinguish the different species with confidence. If you belong to the group that wants to find out more, we hope this book will be exactly what you need. Both authors have spent many years observing and recording dragonflies in Essex and elsewhere, but we have co-ordinated a more systematic survey of the county's dragonflies over the past six years. Many recorders across Essex have supplied us with a great deal of information, and this book is an attempt to summarise this wealth of knowledge and make it available to fellow naturalists.

What is a dragonfly?

The answer might seem obvious – but there are a few complications! The first of these is that we use the word 'dragonfly' with two different meanings. Mostly (and the first paragraph above is a typical example) we use the term to apply to the large, robust and powerfully flying species that readily catch the eye. In fact, these belong to the sub-order 'Anisoptera' within a

Structure of typical dragonfly

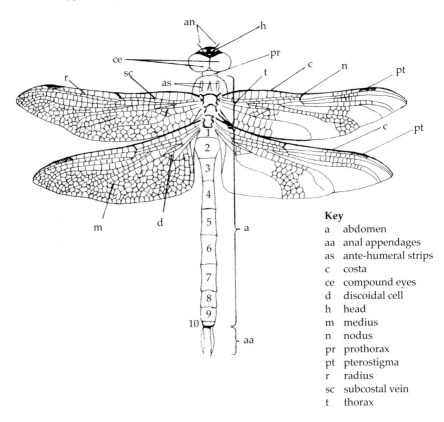

Key
a   abdomen
aa anal appendages
as ante-humeral strips
c   costa
ce compound eyes
d   discoidal cell
h   head
m   medius
n   nodus
pr prothorax
pt pterostigma
r   radius
sc subcostal vein
t   thorax

Leg of dragonfly

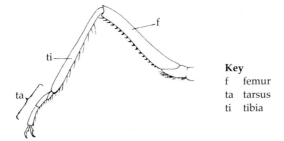

Key
f   femur
ta tarsus
ti tibia

larger grouping (order Odonata) that includes both the dragonflies in this narrow sense and the more delicately built 'damselflies'. These belong to the sub-order 'Zygoptera'. So, sometimes 'dragonflies' can refer to the Anisoptera only, and sometimes to the larger grouping, order Odonata, including both the dragonflies and damselflies. Hopefully the context will make it clear which!

Like other insects, the bodies of dragonflies are segmented, and made up of three main structures: the head, thorax ('chest') and abdomen. The head carries the large, compound eyes, which give them 360° vision. Other sensory organs are also situated on the head – these include three simple eyes (ocelli) on top of the head and a pair of small, rather undeveloped, antennae. At the front of the head are the powerful jaws (mandibles) with which their prey can be held and chewed. The thorax contains the powerful flight muscles, and both the two pairs of wings and the six legs are attached to this part of the body. Unlike most other insects, the muscles that work the wings in the dragonflies are attached to the wing-bases (and not to the outer 'skeleton' of the thorax). This is what gives dragonflies such extraordinary versatility in flight – including the abilities to hover and to fly backwards. Dragonfly wings are membranous, without scales, and their structure is maintained by a fine network of veins. Towards the tip of each wing is a small, dark marking, the 'pterostigma'. The abdomen is usually long and narrow, giving the dragonflies their distinctive appearance (and giving them one of their more colourful common names 'devil's darning needles'). However, some of the medium-sized dragonflies ('chasers' and 'skimmers') may have wider, more 'chunky' abdomens.

The abdomen consists of ten segments, each with a pair of breathing holes, or 'spiracles'. It also contains the rear part of the digestive system, as well as the dragonfly's reproductive organs. In males, there are two pairs of claspers, or 'anal appendages' at the tip of the abdomen. These are used to hold on to the females during mating, and, in some species, while she lays her eggs. There is also a pore on segment nine, towards the tip of the abdomen, from which sperm are emitted, but these are transferred to a structure on segments two and three of the abdomen either before or during mating. The females collect the sperm from this structure, the 'accessory genitalia', and have corresponding appendages at the tip of their abdomen. Also present near the tip of the female abdomen is a device for egg-laying, the 'ovipositor'. This is quite prominent in those species which cut openings into plant tissue to lay their eggs. Species (such as the common darter, *Sympetrum stri-*

olatum) that do not lay their eggs in plant tissue have a 'vulvar scale' on abdominal segment 8. Differences in the structure and orientation of this can be useful in identification of some species.

The two sub-groups of dragonflies (Odonata) differ from each other in several ways. As noted above, the damselflies (Zygoptera) tend to be more delicately built and fly less powerfully than the dragonflies (Anisoptera). There are other differences that help to distinguish the two groups. Both groups have large compound eyes, but these are located differently on the head. In the damselflies the eyes are widely separated from each other, whereas in the Anisoptera the eyes cover most of the surface of the head, and meet, or almost so, over the top of the head. In the damselflies the fore- and hindwings are similar in shape, but in the Anisoptera the hind wings are broader than the forewings. There are behavioural differences, too, but these are more variable within the two groups. In general, the Anisoptera settle with their wings fully spread out, whilst the damselflies usually settle with their wings folded back over their abdomen. However, some of the Anisoptera (especially the medium-sized 'chasers', 'darters' and 'skimmers') tend to point their wings down and slightly forwards when fully at rest. Some damselflies – notably the 'emeralds' (*Lestes* species) – tend to settle with their wings partly open.

Life history

Insects are often divided into two main groups on the basis of their different life histories. Those with 'complete' life histories include flies (Diptera), butterflies and moths (Lepidoptera) and bees, wasps, ants and their relatives (Hymenoptera). In these insects, the life history has several well-defined stages: egg, larva, pupa (or 'chrysalis') and, finally, the adult. For these insects, the larval stage is relatively simple in structure, and the main concern is feeding and growing. When full-grown, the larva sheds its 'skin' for the last time to form the pupa. This is usually referred to as a resting stage as there is little or no movement, and no feeding. However, internally it is anything but resting. It is during this stage that the amazing internal transformations that will lead to the emergence of the adult insect take place.

By contrast, such insects as grasshoppers and crickets (Orthoptera) and dragonflies have so-called 'incomplete' metamorphoses. In these insects, there is no resting, pupal stage. Instead, there is a more gradual transition through a succession of moults from newly hatched larvae (sometimes called nymphs) to the final emergence of the adults. As they lack the pupal stage, the larvae of these insects tend to be more structurally complex, and more closely resemble the adult forms than is the case with the larvae of insects with complete metamorphoses. This is certainly true of dragonflies. Although their larvae are aquatic, and so do not have wings, they can usually be identified down to species on the basis of their external features.

The methods of egg-laying used by dragonflies are very variable. Damselflies generally lay their eggs in the tissue of water plants, as do many of the large 'hawker' dragonflies. However, some of these (notably the migrant hawker, *A. mixta* and the southern hawker, *A. cyanea*) will often lay their eggs in muddy soil or among mosses at the edge of a pond.

In fact, when engaged in this activity, female southern hawkers seem indifferent to the substrate. A female observed by one of us as she laid her eggs around the edge of a pond, continued to attempt to lay eggs on his clothing, spectacles and face before continuing round the pond! The darter (*Sympetrum*), skimmer (*Orthetrum*) and chaser (*Libellula*) dragonflies usually drop their eggs freely into the water, or, sometimes, onto vegetation away from open water. The eggs of species that oviposit in plant tissue tend to be elongated and cylindrical, while those that are dropped freely tend to be more rounded in shape. In most species the eggs develop soon after being laid, and hatch from two to five weeks later. However, in some species, especially those which inhabit pools or ditches liable to drying out in the summer, the hatching of the eggs is delayed. After partial development the embryo enters a resting stage ('diapause'), until a combination of rising temperature and moisture levels triggers hatching the following spring.

The form that emerges from the egg is known as a 'prolarva'. It has specialised adaptations for breaking free from the egg and surrounding material, but also, where necessary, for reaching the water. Prolarvae that hatch away from the water may simply drop into it, or may have to find water by a series of jumps. These may cover a distance of several centimetres! Depending on species, the prolarval phase is very brief, lasting from a few seconds to, in a few cases, several hours. After this time the insect moults, and takes on the more typical larval form of its species.

There are very marked differences in bodily form between the larvae of the damselflies and those of the dragonflies (Anisoptera). The former are generally elongated and, like the adults, rather delicately built. Also like the adults their bodies are divided into head, thorax and abdomen, with six legs, as well as functionless wing-stubs, attached to the thorax. There are also two striking structural adaptations to their aquatic way of life. The first is the 'mask', a complicated structure formed from the insect's mouth-parts. This is generally kept folded away under the larva's body, but is suddenly unfurled forwards to catch its unfortunate prey. The other feature is at the tip of the abdomen. It consists of three wide, flattened plates, or 'lamellae'. These are supplied with a network of fine veins, and they are the larva's breathing apparatus. Their shape and structure are often useful in identifying damselfly larvae.

The larvae of the dragonflies (Anisoptera) are, like the adults, more sturdily built. They share with the damselflies the distinctive 'mask' mechanism for catching their prey, but their breathing apparatus is different. Instead of the caudal lamellae they have a hollowed out rectum, with a greatly folded interior surface. Water is pumped into and out of the abdomen, so that oxygen and carbon dioxide are exchanged across the rectal surface. This mechanism is also used in locomotion, as a kind of jet propulsion.

The larvae have different modes of life according to species. Some live partly submerged in the muddy bottom of ponds or slow-moving rivers (*Libellula, Orthetrum*, for example), while others live on the bottom, and are usually camouflaged dark brown in colour (*Calopteryx, Platycnemis, Pyrrhosoma*), while still others live among water weeds (*Anax*, some of the *Aeshnas, Coenagrion, Sympetrum*, for example). The latter group tend to be greenish or yellowish in colour with variable darker mottling, and some can change colour when they moult. The larvae of both damselflies and dragonflies are fiercely predatory,

Female migrant hawker egg-laying

and use their camouflage to lie in wait for passing prey. They are usually non-specific in their choice of prey: from single-celled animals, water-fleas and tiny worms in the case of smaller larvae, through to small fish, beetle larvae, tadpoles and other dragonfly larvae, in the case of the full-grown

Anisopteran larvae. Usually it is the movement of the potential prey that stimulates the larva to shoot out its 'mask' and grab its victim. This requires great precision as well as speed, and in most species it seems that the victim is first pin-pointed with a stare of the compound eyes.

Development through the larval stage is very variable, depending both on the availability of food and the surrounding temperature. Each species also has its own seasonality and developmental rhythm. In some species, such as *Lestes sponsa*, larval development is very rapid, being completed in as little as two months. This, taken together with the prolonged 'resting' phase

A damselfly larva

(diapause) of the egg, gives an annual life cycle. However, many species take two or more years to complete their full life cycle, some as long as five or more years. In general, the bottom- or mud-dwellers tend to develop more slowly as the surrounding water temperature tends to be cooler, while the weed-dwellers tend to develop relatively quickly. It seems that larger dragonfly (Anisopteran) larvae are better at surviving cold winter temperatures than smaller ones, so that eggs laid by late-flying species tend to pass the winter in that stage, whereas those laid by spring-flying species will hatch in the normal way. The resulting larvae can then develop through the warmer seasons and be ready to survive the winter cold.

As with other insects, growth can only occur by repeated shedding of the tough, inelastic outer skin or 'cuticle' of the dragonfly larva. The number of moults required for full development varies, according to species and conditions, from eight to eighteen. At each moult the larva expands its thoracic area until the cuticle splits open. The larva then emerges from its former skin, and rests for a while until its new cuticle hardens. When fully grown, the larva is ready to undergo its final transformation, but this does not necessarily happen right away. It is important that adult male and female dragonflies are on the wing at the same time, otherwise mating and reproduction could not occur. So, the emergence-periods of dragonflies tend to be closely regulated. This is especially true of those species that fly in the spring. In these species the flight-period is relatively short, and emergence from the larval stage is synchronised. Other species (the common darter dragonfly (*S. striolatum*) and common blue damselfly (*E. cyathigerum*) are good examples) ensure that males and females are on the wing at the same time by having a more prolonged emergence and flight-period.

Just before its final moult the larva comes closer to the surface of the water, or migrates to the shallow margins of its habitat. Some internal changes can be seen through the larval cuticle at this stage. When conditions are suitable for emergence, the larva leaves the water, often using the rigid stem or leaf blade of an emergent plant such as reedmace or iris. Once out of the water, the larva will sometimes climb some distance up its support before securely positioning itself. In many species the larva will wriggle its abdomen from side to side, possibly to ascertain that there is enough room around it for the uninterrupted expansion of its wings. The next phase resembles a normal moult, with expansion in the area of the head and thorax causing the cuticle to split open. In the case of the larger Anisoptera, the adult dragonfly forces its way out through this split and then hangs motionless below its cast-off skin, attached only by the hind segments of its abdomen. After resting for a while it bends its body forwards again,

to grip the plant stem, and release the remaining abdominal segments. Soon afterwards it 'pumps up' its wings to their full size. The cast-off larval skin, or 'exuvia' can remain in place for some time and provides a useful way of counting the numbers of dragonflies of a species emerging from a site.

As soon as the new cuticle has hardened sufficiently, the dragonfly is able to make its first flight. At this point the body is pale and lacks the full colouration typical of its species, and the wings are quite opaque. Dragonflies of most species disperse away from the breeding sites during this, very vulnerable, 'teneral' phase of their

Newly emerged emperor dragonfly with the larval skin (exuvia)

development. Newly emerged dragonflies are particularly vulnerable to predation by birds, and this may explain why many species emerge during the night and take their maiden flight very early in the morning. During the teneral phase they can often be seen hunting for prey in sheltered woodland rides or glades, or in rough grass or heathland well away from water. When fully adult and ready to mate, they may return to their original habitat or seek out suitable breeding habitat elsewhere.

Adult behaviour

Adult dragonflies continue the predatory ways of their earlier larval existence. Of course, since they have now become inhabitants of the air and dry land, the types of prey and methods of capture are very different. Instead of lying in wait, adult dragonflies and damselflies actively hunt for their prey.

Male emperor with migrant hawker as its prey

Hairy dragonfly eating a captured bumblebee

This is most conspicuous in the case of the large 'hawker' dragonflies. These patrol woodland rides, pond or river banks and other insect-rich habitats, swooping to catch their prey in flight with their mouth-parts, and trapping it in a 'cage' with their legs. Often the prey will be small, flying insects such as midges or winged ants and the hawker will catch and consume them without interrupting its flight. However, larger prey such as beetles, bumblebees, butterflies or even other, smaller hawkers may cause the hawker to come to rest and gnaw away at its victim among rough vegetation. One of us recently observed an emperor catch a male migrant hawker in flight, settle with it in a tussock of grass, bite off its head, and work its way through the thorax and upper abdomen. Meanwhile the body of the luckless victim continued to squirm.

The medium-sized 'skimmer', 'darter' and 'chaser' dragonflies may also be seen hunting for prey in a similar manner, but they also spend much of their time settled on prominent 'perches' (or, in some cases, on bare ground, or stones). From their perches they fly out on short sorties to catch passing prey.

Common blue damselfly eating
a small moth

A male four-spotted
chaser on its 'perch'

Males also use their perches as look-out posts to secure their territories at breeding sites, and to chase after passing females. After each sortie they return to the same perch, or one of a small number of adjacent favourites. The damselflies also hunt for their prey, usually other small insects such as micro moths, gnats, and other, smaller, damselflies.

At the breeding sites there is frequently aggressive interaction between males of most species. This may be directed at rivals of the same species, or individuals of either sex of other species. The large, powerful male hawkers will often patrol a section of water-margin for hours on end, chasing off other dragonflies that they encounter – other hawkers, or smaller darter or chaser dragonflies. Particular aggression is elicited by intruder males of the same species, resulting in spectacular displays of aerial combat, and the eventual retreat of one of the combatants. Both males and females hawk along banks or ditches for prey, but males are also on the lookout for newly emerged females. However, they will also dive down and attempt to mate with females as they are occupied with laying eggs. If a male is successful in mating with an already-mated female he will first clear out the sperm of the earlier mate.

Not all Anisopteran males are so strongly territorial, however. Males of the migrant hawker (*A. mixta*), for example, often congregate at rich feeding sites, which may be well away from water, in large numbers. However, when patrolling at breeding sites they do show more inclination to cover a selected section of water margin and there are brief conflicts between males that stray into one-another's territory, but these are mild and short-lived interactions compared with those shown by, for example, the emperor. Also, several patrolling males of *A. mixta* may co-exist along the margins of a single small pond – something rarely tolerated by male emperors. Aggressive behaviour in damselflies is not so spectacularly visible, but it is certainly present. Rival males will often 'buzz' mating or egg-laying pairs, and there is considerable competitive interaction between damselflies of different species. Males of the common blue damselfly (*E. cyathigerum*)

A male downy emerald hovering in search of a female

seem especially aggressive, and will attack males of the common red-eyed damselfly (*E. najas*) that tend to share the same perches on floating leaves. Where several damselfly species are present at a pond there is usually a discernable spatial partitioning between the different species, presumably produced by aggressive interactions. One of us watched a male *E. cyathigerum* pick up a female *L. sponsa* from its perch near the centre of a pond and fly with it to the margin. After being dropped in the water the chastened *sponsa* quickly climbed up a plant stem and waited for its wings to dry.

Another characteristic behaviour shown by males of some species (for example, migrant hawker (*A. mixta*) and downy emerald (*C. aenea*)) at the breeding sites is intermittent hovering during a patrol. The insect stares fixedly for several seconds at a small section of emergent vegetation at the water margin, before moving off again, often contouring the interface of marginal vegetation and the open water, but also threading its way through plant stems and occasionally flying a metre or so over the bank. The hovering behaviour is presumably a way of visually fixing on any newly emerged females that might be clinging to emergent vegetation.

In most species there is no evidence of 'courtship' behaviour, although in some Anisoptera females unwilling to mate signal this to males by bending their abdomen downwards. In most damselflies unreadiness to mate is signified by a rapid flicking opening and closing of the wings. However, the damselflies of the genus *Calopteryx* (the banded and beautiful demoiselles) and the white-legged damselfly (*P. pennipes*) do have noticeable courtship displays. In the demoiselles, the male signals to a passing female by rapidly opening and closing his wings. If prepared to mate she settles nearby, while the male flutters spectacularly over and around her. The sight of large numbers of males engaged in this activity on a sunny day is one of the great thrills of dragonfly-watching. In the courtship of the white-legged damselfly the male exhibits his distinctively coloured legs to his chosen female.

Both dragonflies (Anisoptera) and damselflies have a unique method of mating. The following sequence gives the general pattern for both damselflies and dragonflies, though the details vary somewhat from species to species. The male detects the female mainly by visual appearance. If

A male beautiful demoiselle 'flashing' at a passing female

not rejected by her, he lands on her back and curves his abdomen under and below his body so as to grasp the female in the region of the junction between her head and thorax with the claspers at the tip of his abdomen. She then curves her abdomen below and forwards, in the same way, so that the tip of her abdomen connects with the 'accessory genitalia' on segments two and three of the male's abdomen. The male will have already transferred sperm from his genitalia in segment nine to his accessory genitalia, where they are transferred to the female.

Mating pair of migrant hawkers

This elaborate posture, reminiscent of an illustration from the Khama Sutra, is generally known as the 'wheel' position. Some species mate in flight, and the female begins egg-laying almost immediately. More usually, even where mating is commenced in flight, the pair will soon settle in nearby vegetation. They may remain in the wheel position for a few minutes, or up to half-an-hour, depending on species. If disturbed they usually remain coupled and fly off to find an alternative hiding place.

Mating pair of blue-tailed damselflies

Since most breeding sites are highly competitive places, and dragonflies are capable of mating many times, males of most species

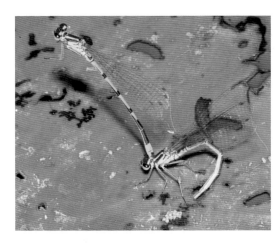

A pair of azure damselflies
egg-laying 'in tandem'

have ways of ensuring that the female whose eggs they have just fertilised will go on to lay them without interruption by competitors. Some, such as the males of the chasers (*Libellula* species), will often fly close, keeping guard over the female as she repeatedly dips the tip of her abdomen into the water to release her eggs. Others, such as the common darter (*S. striolatum*) often stay connected together in what is known as the 'tandem' position. The male, flying almost vertically with the female hanging below, takes the lead in maintaining a rhythmic bobbing and sweeping motion above the water, swiping the tip of the female's abdomen repeatedly into the water. With each dip, eggs are released into the water. Most of the damselflies (but not *Calopteryx*) usually lay their eggs in tandem. However, instead of releasing their eggs into the water, the females cut slits in water plants and insert their eggs into the plant tissue. Often the female clings to floating vegetation and curves her abdomen below the water surface to lay her eggs, while the male remains motionless, almost upright above her, with legs held tightly against his body.

Pair of common emerald
damselflies egg-laying

However, it is not uncommon for both members of the tandem pair to make their way down a plant stem until both are completely submerged. At the opposite extreme, emerald damselflies (genus *Lestes*) often lay their eggs in the stems of reeds or rushes a metre or more above the surface of the water (or damp mud, as their habitats are often ditches or shallow ponds that dry out in the summer).

Habitats and conservation

All British dragonfly species spend their early lives as aquatic larvae, and their adult lives in the air or on land. They are therefore dependent both on the quality of the water-bodies in which they breed, and on the surrounding terrestrial habitat in which they shelter during vulnerable periods in their life history, and from which they gain their food. More than this, they require suitable transitional zones between water and land to facilitate emergence from the larval stage, as well as egg-laying and mate-location. It is arguable that the most decisive and vulnerable parts of a dragonfly's life take place within a metre or so either side of the water margin.

But beyond these more general comments, we need to consider the variations in dragonfly modes of life. Most obviously, aquatic habitats themselves vary in many respects: still or moving water, swiftly or slow-moving, shallow or deep, alkaline or acidic, shaded or open to the sun, warm or cool, well-vegetated or bare, nutrient-rich or nutrient poor, fresh or saline and so on. The numbers and species-mix of any site will depend on the peculiar local combination of these and other factors.

A useful way of capturing some of the patterns is to distinguish between still- and moving-water habitats, and, among the latter, between swiftly flowing streams and rivers, and more slowly-flowing lower reaches of rivers. Among the still-water habitats we can include long-established farm and village ponds as well as older mineral-extraction sites. These tend to be small and have shallow, well-vegetated margins. Similar conditions can often be provided in garden ponds, and artificial ponds created in reserves and country parks specifically for conservation purposes. Depending on the quality of the surrounding habitat, such ponds can sustain most of our common dragonfly and damselfly species. Larger lakes were often associated with the landscaped parks of country houses, and many remain as features in local authority-run country parks. Other large bodies of water have been formed as water reservoirs, or as a result of modern mechanised sand-and-

*Black-tailed skimmer
sun-bathing*

gravel extraction and farm reservoirs. Such sites are very variable in their dragonfly-friendliness: where they are very heavily stocked with fish for commercial angling, or have hard, un-vegetated margins, they are likely to be used by relatively few species. Where lakes are newly formed by flooding of mineral extraction sites it is an instructive study to observe their colonisation by successions of species as both aquatic and marginal vegetation develops. Often the common blue damselfly (*E. cyathigerum*), broad-bodied chaser (*L. depressa*) and the black-tailed skimmer (*O. cancellatum*) will be among the first arrivals (and they will also thrive in hard- and bare-edged older lakes and ponds). None of these species seems to require emergent or marginal vegetation and the skimmer is particularly partial to basking on patches of bare ground on or near the bank.

Two more specialised types of still-water habitats are important for populations of some of our more scarce or localised species. These are (usually acidic) pools in heaths, mosses and bogs, and ditches and dykes in low-lying fenland or grazing marshes. The remaining extensive heathlands of Dorset, Hampshire and Surrey are the southern English strongholds of several localised species such as the black darter (*S. danae*), common hawker (*A. juncea*), downy (*C. aenea*) and brilliant emerald (*S. metallica*) dragonflies, keeled skimmer (*O. coerulescens*) and small red damselfly (*C. tenellum*), together with many of the more widespread species. Because of intense development pressures and conflicting management priorities these are among the most threatened of all wildlife habitats. East Anglia (including coastal and estuarine districts of Essex) has been well-supplied with low-lying fens and marshes, though most of our former wealth of wetland habitat has long since been replaced by arable agriculture. However, the Norfolk Broads, some Suffolk coastal areas and the remnants of the Essex marshes still harbour important populations of such species as the variable damselfly (*C. pulchellum*), the hairy dragonfly (*B. pratense*), the four-spotted chaser (*L. quadrimaculata*), the migrant hawker (*A. mixta*), the ruddy darter (*S. sanguineum*) and both the common and scarce emerald (*L. sponsa* and *L. dryas*) damselflies. The Norfolk hawker (*A. isosceles*) is another species of this habitat, but with a much more restricted East Anglian distribution than the others.

Turning, now, to flowing water habitats – rivers and streams – we find different assemblages of characteristic species. In upland areas of northern and western Britain, as well as in the southern heath districts, the magnificent golden-ringed dragonfly (*C. boltonii*) breeds in small streams and seepages. Elsewhere, more characteristic species of swiftly flowing streams and rivers are the large red damselfly (*P. nymphula*) and the aptly named beautiful demoiselle (*C. virgo*), often flying together with the ubiquitous common blue-tailed damselfly (*I. elegans*). As we move along to the lower, more slowly flowing reaches, especially where the aquatic and marginal vegetation is well-developed, the beautiful demoiselle is joined by, and eventually replaced by, the banded demoiselle (*C. splendens*), the common blue (*E. cyathigerum*) and azure damselflies (*C. puella*), the brown, migrant and southern hawkers (*A. grandis, cyanea* and *mixta*), the emperor (*A. imperator*), and, more locally, the hairy dragonfly (*B. pratense*), scarce chaser (*L. fulva*), the red-eyed damselfly (*E. najas*) and white-legged damselfly (*P. pennipes*). Many of the species that breed in slowly flowing lower reaches of rivers also breed frequently in still-water habitats, but some, such as the scarce chaser and white-legged damselfly, are much less likely to use still-water sites.

Although dragonflies are themselves fierce predators, they provide an attractive meal for other predatory animals, and also suffer from a range of parasites. Dragonfly eggs are parasitized by a tiny wasp, and the larvae by a Trematode worm that completes its life history only if the larva is later eaten by a bird. Parasitic water mites also infect large numbers of larvae, and are dispersed by being carried on the body of the freshly emerged host. Smaller dragonfly and damselfly larvae are often eaten by larger ones, but they are also preyed on by a range of other aquatic predators, such as water beetles, water bugs ('water boatmen'), larger fish, amphibians and also water birds such as moorhens, ducks and herons. As mentioned above, the adult dragonflies are at their most vulnerable just after emergence and during the following week or so when there are still in the immature 'teneral' phase. Corbet observed the emergence of emperor dragonflies at a pond, over two successive years. These were preyed on by a pair of blackbirds in early mornings, but this predation was a relatively small (3–5% of the total emergence) cause of mortality compared to other factors. For example, many of the emerging dragonflies failed to fully expand their wings and fell back into the water, only to be consumed by waiting newts (Corbet, Longfield & Moore 1960).

Of course, many other bird species prey on dragonflies. One of us once observed a scarce chaser begin its maiden flight only to be intercepted on its way up to nearby tree canopy by a spotted flycatcher. Hobbies are also noted dragonfly predators. Among the more significant invertebrate predators on adult dragonflies are spiders of several species. Webs of *Tetragnatha extensa* are often festooned with the remains of damselflies, and the recent UK

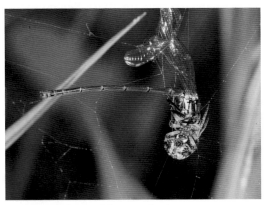

colonist, the wasp spider (*Argiope bruennichi*), must also account for many. In a pond we regularly visit, webs of this species are now strategically placed among grass tussocks around the pond-margin.

There are several accounts of vespid wasps as predators on dragonflies (e.g. Corbet 1999, Benton 1988) and an interesting description of hornets (*Vespa crabro*) attacking numerous *A.*

Large red damselfly caught in a spider's web

mixta along a lake margin (Cham 2004). Cham's observations suggest that the success rate of daytime predation is low, but he watched a hornet successfully attack and 'butcher' a roosting *mixta* in early morning. The hornet divided up the dragonfly's body and took it away (presumably to its nest) section by section. Insectivorous plants, too, capture damselflies, and even hawker dragonflies. One of us watched a luckless common hawker (*A. juncea*) inextricably caught in a patch of sundew in a Dorset bog.

However, by far the greatest threats faced by dragonflies are changes to their habitats: either the result of natural succession, or caused by human

intervention. Most drag-
onflies, especially the ones
that breed in still-water
sites, are good at dispersing and discovering new
breeding sites. So long as
new breeding sites are being formed, within dispersal range, natural succession is no great threat. In
Essex, flooded pits formed
by sand and gravel extraction are a major dragonfly
habitat (breeding habitat for
some twenty species), and
are quickly colonised by a
range of dragonfly species.

*A common hawker caught by
sundew plants*

As shallower pits become
dryer and choked with emergent plants such as reedmace or rushes they are
particularly favoured by the emerald damselflies (*Lestes* species), while early colonisers have dispersed to other, more recently formed pits. The main
threats to this type of habitat are various sorts of human activity. Landfill
and restoration to arable agriculture are the most obvious of these, but the
use of many former quarries for angling has mixed consequences. At some
sites it seems that angling can coexist with a rich dragonfly population, but
over-stocking with fish can certainly be damaging. Detailed research on this
would be well worth doing.

Apart from flooded pits, farm and village ponds, garden ponds and ornamental lakes provide variable resources for many species of dragonfly. With
reduced outdoor grazing, and the introduction of piped water supplies,
farm ponds have either disappeared, or become too overgrown and polluted to have much value as dragonfly habitat. In many cases, such relict
ponds are also shaded over by clumps of surrounding trees. Many village
ponds and ornamental ponds and lakes are adversely affected by eutrophication from water-fowl droppings. The result is often a loss of the structural
supports and hiding places supplied by the larger-leaved aquatic plants as
well as the associated invertebrate food sources that dragonfly larvae need.
Shading out by overhanging trees, and eutrophication due to rotting leaves
is another common feature of many such sites. To encourage strong popula-

tions of the commoner species, ponds should ideally be in sheltered spots, with some trees or shrubs nearby. However, shading by overhanging trees should be avoided, especially on the southern margins, so that the pond is open to the sun. The water should be shallow at the margins, with a band of emergent and marginal vegetation around the edges, and with both submerged aquatic plants such as hornwort or Canadian pondweed, and patches of floating-leaved plants such as water lily or pondweed (*Potamogeton* species). A hinterland of open, invertebrate-rich grass or heathland should be accessible by adult dragonflies – in the shape of wide rides and glades if the pond is situated in woodland.

Ditches and dykes rarely support large numbers of dragonfly species, but they are important habitats for a small number of more localised species. There are four main threats to these habitats. Where they run through or alongside arable cultivation they are liable to pollution from agricultural chemicals, fertilisers or pesticides, or both. A second problem affects ditches adjacent to the sea walls. These harbour some of the more localised species such as the scarce emerald damselfly (*Lestes dryas*) and the hairy dragonfly (*Brachytron pratense*). Both species seem tolerant to a degree of salinity, but are unable to survive where there is regular leakage of sea water into the ditches. The other two threats are from under-, or over-zealous management. Mechanical clearing and deepening of the ditches makes them unsuitable for most species until vegetation re-colonises, while without management, successional changes lead to ditches becoming choked with vegetation and eventually drying out. Neither of these processes is necessarily harmful in the long run, so long as there is sufficient diversity in timing of management activity in a local area to allow for a variety of different successional stages to co-exist within the dispersal ranges of the dragonflies.

In general, moving-water habitats have tended to fare rather better than some other types of habitat. One major threat is from pollution incidents up-stream. Release of raw sewage, and escape of toxic chemicals from road accidents have combined with more sustained leakage of agricultural run-off to severely damage the aquatic life of several Essex river systems in recent years. However, the dragonfly fauna has generally recovered quite well from such incidents, and several of the Essex rivers now support a richer dragonfly fauna than was present in the 1980s. Factors that have been shown to affect the presence and physical distribution of dragonfly breeding habitat along river systems have been shown to include rate of flow, shade, cover provided by aquatic plants and concentrations of oxygen and phosphates.

	Lakes and ponds	Flooded pits	Coastal and estuarine ditches	Slow rivers and canals	Swift rivers and streams
C. virgo					+
C. splendens	(+)			+	+
L. sponsa	+	+	+	(+)	
L. dryas	(+)	(+)	+		
P. pennipes	(+)			+	
P. nymphula	+	+	(+)	+	+
I. elegans	+	+	+	+	+
C. puella	+	+	(+)	+	+
E. cyathigerum	+	+	+	+	
E. najas	+	+		+	
E. viridulum	+	+		+	
B. pratense	(+)		+	+	
A. cyanea	+	+	(+)	(+)	
A. grandis	+	+		+	
A. mixta	+	+		+	+
A. imperator	+	+	+	+	
A. parthenope	+	+			
C. aenea	+	+			
L. depressa	+	+			
L. quadrimaculata	+	+	+	+	
L. fulva	+			+	
O. cancellatum	+	+	+	+	
S. sanguineum	+	+	+	+	+
S. striolatum	+	+	+	+	+
S. flaveolum	+		+		
S. fonscolombii	+	+			

Table 1 Species recorded at different habitat types during the present survey. Brackets indicate that records have been obtained from an unusual habitat for the species

In general, upper reaches, with faster flow, are tolerated by rather few species, with greater species diversity in slower-flowing, and usually more vegetated downstream reaches. However, this pattern is affected by management regimes and patterns of human river use. River-straightening, dredging and bank clearance to reduce flooding risk affect conditions for larval development as well as habitats for oviposition and emergence at the river margin. Features such as weirs also affect both rates of flow and oxygen content. Use of rivers for boating may have some effect through disturbance of the river-bed, and presumably this is intensified with regular motorised boat traffic. The Chelmer-Blackwater complex still has one of the richest dragonfly assemblages in the county, but is increasingly used by motorised river transport. The effect of this source of disturbance and pollution on dragonflies and other aquatic invertebrates is in need of serious study. Since adult dragonflies also require nearby habitat for feeding, mating and roosting, the surrounding terrestrial habitat significantly affects their survival. In general, arable agriculture bordering the river is less favourable than permanent pasture or land managed for amenity (see Hofman & Mason 2005, on the River Stour).

The dragonflies of Essex

According to our survey, Essex currently has 23 established breeding species. This compares with the current total of 38 established breeding species in the UK as a whole. However, with increasingly frequent sightings of several species that breed close to us on the European mainland, it seems likely that these totals will soon be increased. The recent colonisation and rapid spread of the small red-eyed damselfly (*E. viridulum*) seems likely to be the first of several new additions to our dragonfly fauna.

Of the dragonfly habitats discussed above, Essex is least well-endowed with heathland pools and acid bogs. Historically, the Epping district had some extensive areas of wet heathland, and Edward and Henry Doubleday reported the presence of several of the characteristic dragonfly species of this habitat in the 19th century. These included the small red damselfly (*C. tenellum*), the black darter (*S. danae*) and scarce blue-tailed damselfly (*I. pumilio*). There is also an early 20th century record of the keeled skimmer (*O. coerulescens*) from the same district (see **Chapter 4**). However, these habitats were gradually lost through drainage and succession, with the black darter surviving in Epping Forest until at least the late 1940s. In recent years, conservation management has opened up some of the former heathland areas, and there are very occasional sightings of individuals of both the black darter and keeled skimmer. Although it seems unlikely that either species currently breeds in the forest, these sightings do suggest the possibility of future re-colonisation, given appropriate management activity.

By the period (1980–87) of the previous county-wide survey (Benton 1988) there were 21 confirmed breeding species in the county. A further species, the variable damselfly (*C. pulchellum*), had been recorded during that survey period, and was included as a possible breeding species (at one site only). A sighting of the black darter in 1987 also opened speculation that a small breeding population of that species might have persisted in the vicinity of Epping Forest. However, between that time and the current survey (2000–2006) we have only a few, scattered reports of the black darter in the county. Although there are occasional reports of the variable damselfly from the Old River Lea/Cornmill Stream dragonfly reserve, we have been unable to confirm the presence of the species there despite several searches. Further survey work will be required to settle this puzzle.

Otherwise, the 2000–2006 survey confirmed the continuing presence in the county of all the breeding species found in the previous survey. However, there are several significant changes (see Table 2).

Perhaps the most significant of these was the addition of two new species. The first discovery (and probably initial colonisation) of the small red damselfly (*E. viridulum*) in Britain was at a pond in east Essex on 17th July 1999 (Dewick & Gerussi 2000 – reprinted as **Appendix 3**). During the period of the current survey this species has gone on to spread rapidly across the county, and to other areas in East Anglia and south-east England. The second 'new' species was in reality the return, after a century of absence, of the very localised scarce chaser (*L. fulva*). Doubleday reported it as 'rare' (Doubleday 1871), and there is just one subsequent report of it, by Harwood, probably in 1900, on the Colne at Colchester (Harwood 1903). Amid signs of an expansion of range in other parts of Britain, it was reported from a site in south-west Essex in 1997 (T. Gunton, pers. corr.), and also in that year first reported from the Stour near Nayland. Again, during the period of the current survey this species has greatly increased its population along the Stour, reappeared at several locations along the River Colne, and established itself on the Chelmer/Blackwater complex. The continued presence of both new species in the county seems assured.

The discovery of the scarce chaser along the Stour coincided with what appears to be a marked recovery of the dragonfly fauna of the river from the mid-1990s onwards. The previous survey concluded that the river 'seems to have formerly supported a rich Odonata fauna, including several rare or local species. It now supports very few species (along its Essex reaches,

	1980–87	2000–06
C. virgo	4	1
C. splendens	30	42
L. sponsa	30	33
L. dryas	10	16
P. pennipes	6	19
P. nymphula	24	43
I. elegans	56	56
C. puella	43	53
C. pulchellum	1	
E. cyathigerum	54	56
E. najas	13	43
E. viridulum		39
B. pratense	1	21
A. cyanea	46	51
A. grandis	46	51
A. mixta	51	57
A. imperator	31	54
A. parthenope		2
C. aenea	2	4
L. depressa	27	47
L. fulva		5
L. quadrimaculata	14	36
O. cancellatum	24	50
S. sanguineum	42	52
S. striolatum	54	55
S. flaveolum		2
S. fonscolombii		7

Table 2 Comparison of two surveys of Essex dragonflies
(1980–7 and 2000–6) by 10km squares

at least) and these are often present at rather low densities' (Benton 1988: 32–3). The current situation is entirely different. In addition to the scarce chaser, the red-eyed (*E. najas*) and white-legged (*P. pennipes*) damselflies are now abundant in many places, and most of the more widespread species have also been recorded. A recent academic study yielded 15 species of adult dragonfly along the river, and ten species of larvae, mostly from the lower reaches (Hofmann & Mason 2005). The explanation of this increase in the dragonfly fauna of the Stour remains an open question. One suggestion is that large volumes of water are now pumped underground from Denver on the Ely Ouse into the upper reaches of the Stour (Cham 2000). This water could contain eggs or larvae, including those of the scarce chaser, which does occur in the Ouse Washes. However, it cannot be ruled out that a small population survived on the Stour, undetected, from former times, or that, given reported expansions elsewhere, the species established itself naturally.

As well as these additions to the Essex dragonfly list, it is also encouraging that several species reported as scarce or very localised in the previous survey seem to have significantly improved their status in the county since then. The white-legged damselfly (*P. pennipes*) was recorded from only three river systems in Essex during the 1980s, was presumed lost on the Stour, and in decline on the Chelmer/Blackwater. The present survey shows a marked recovery. It now has a strong population on the Chelmer/Blackwater, and is present on several of its tributaries. It may have been present but overlooked on the Stour in the earlier survey, or may have colonised from higher reaches of the river in Suffolk. However, the recovery of the species on the Essex/Suffolk stretches of this river is quite dramatic, and coincides with the re-colonisation and flourishing of the scarce chaser. The large red damselfly (*P. nymphula*), too, has been reported from many more sites in the current survey, including several in the north-west of the county, from which it had appeared to be absent. This species has a rather short, early flight period, and it seems likely that some of the increase between the surveys is an effect of more thorough surveying in the later. The red-eyed damselfly (*E. najas*) was formerly very localised in the county, with a mainly western and north-eastern distribution. Despite careful surveying it was not found on the Chelmer/Blackwater and was thought to have been lost from that locality. However, the current survey shows a much more optimistic picture, with a greatly increased spread and density of sites across the county. It is now common along the Chelmer/Blackwater and on several of its tributaries, as well as occurring on other rivers and numerous still-water sites. The national range is mainly southern, with evidence of a northward expansion in the past two decades (Brooks 2004).

Another encouraging example is that of the hairy dragonfly (*B. pratense*). There are historical records from Epping Forest, Benfleet, Langenhoe marsh and the Chelmer/Blackwater navigation. However, despite thorough searches in the 1980s it could be found only on MoD land at Langenhoe marsh. It was then considered our rarest dragonfly (Benton 1988), with only a precarious remaining 'foothold' in the county. However, from the late 1980s onwards, there were reports of the reappearance of this species in several parts of the county. This recovery coincided with reports of recovery in other parts of Britain – mainly in the South-east and Midlands. The current survey has confirmed the continuing expansion and consolidation of this spread in Essex since 2000. The emperor (*A. imperator*) was reported to be fairly widespread, but local within its range, during the 1980s. There were no records from the north-west of the county. We now have records covering the whole county, and from many more individual sites. To some extent this may be due to more thorough survey work in the current survey, but such a familiar and eye-catching species is unlikely to have been much overlooked previously.

The broad-bodied chaser (*L. depressa*) is another species that appears to have gained ground in the county since the earlier survey, but, as an early species with a relatively short flight period, this may be due in part to less thorough survey work in the previous period. However, its relative, the four-spotted chaser (*L. quadrimaculata*) almost certainly has greatly increased in the county since the 1980s. It was then known from only 14 sites in ten 10km squares, and there was some doubt about its status as a breeding species at some of these localities, given its migratory habits. Now the species is widespread and common throughout the county, except for the north-west. The black-tailed skimmer (*O. cancellatum*) has also become decidedly more common and widespread in Essex since the previous survey. It became established in Essex, in the Epping Forest area, in the late 1930s, and subsequently spread to the rest of the county. Here and in other parts of south-east England, it was probably aided by its ability to colonise newly formed pools in mineral extraction pits. It now breeds in a wide range of still water habitats, as well as in slow-flowing reaches of some of our rivers, and the Chelmer/Blackwater navigation. Finally, the ruddy darter (*S. sanguineum*) had suffered a major decline nationally up to the mid-1970s, but by the period of our previous survey it had recovered well in Essex. However, it remained more localised than the common darter, and its recorded distribution had a coastal bias, suggesting the possibility that immigrant specimens may have exaggerated its status. However, our current survey suggests a continued strengthening of its position across the county, and our subjective impression is that it is now rather more frequent than the common darter (*S. striolatum*), at least in the east of the county.

Two other Essex species are of note. The beautiful demoiselle (*C. virgo*) has one of its most easterly British breeding sites in Essex. This is the Roman River, to the south of Colchester, where it was observed by W.H. Harwood more than a century ago (Harwood 1903). There were a few reports of this species at other sites during the 1980s, but no confirmation of it as a breeding species anywhere but the Roman River valley. During the current survey we have received records from the Roman River only, and it retains its strong population there. Please note, however, that just as this book was going to press, we received an encouraging report of a new site for the beautiful demoiselle (*C. virgo*) – only the second in the county. Nine males and twelve females were observed by Mr. N.M. Rayner in a tributary to the River Colne at TL9427 and TL9428, near West Bergholt on 15th July 2006. The same observer also reports the white-legged damselfly (*P. pennipes*) on the River Colne at TL944273 on 18th July 2006. This is the first report known to us of this damselfly on the Colne.

The other notable species is the downy emerald (*C. aenea*). There are historical records from Hatfield Forest, but in recent times it has been confined to several ponds in Epping Forest. During the current survey its continued presence at a number of Forest ponds was confirmed, and we also have a single record from near Stapleford Abbotts. However, it seems likely that this was a 'wanderer' from the main population in Epping Forest.

As a coastal county, Essex has at least its fair share of migrant species. Of the migrant 'darter' (*Sympetrum*) species, the rarest is the vagrant darter (*S. vulgatum*). One specimen was recorded in Epping Forest in 1906, and another at Colne Point (N. Cuming, pers. corr.) during the 'wave' of dragonfly immigration that took place in 1995. The yellow-winged darter (*S. flaveolum*) has been a more frequent visitor, with early records relating mainly to Epping Forest (possibly because that is where the observers were!), and a small scattering of records from the east of the county. There were two more sightings in Epping Forest during the 1980s, but many more, from numerous sites around the county, in 1995, and one in 1996. During the present survey, we have one report from Walton in 2004 (C. Balchin & J. Rowland) and a further 3 reports so far for 2006. The third migratory *Sympetrum* species is the red-veined darter (*S. fonscolombii*). Although a small number of this species arrived on the east coast of Britain in 1995, we do not have any records of Essex sightings. However, during the current survey we have a record from 2002 (M. Heywood), and sporadic reports in 2003 and 2004. However, during the 'invasion year' of 2006 there were reports from several sites across the county, including strong evidence of breeding success at one. The lesser emperor (*A. parthenope*) is a rare migrant, first recorded in Britain in 1996. We have two confirmed sightings during the current survey (M. Telfer at Bradwell in 2000 and Colin Jupp at Bedfords Park in August 2003). So far as we know, these are the first Essex records.

In the absence of very obvious changes in habitat, it may be that climate change is part of the explanation for the generally improved status of the Essex dragonfly fauna, the newly established species and the increase in sightings of migrant species. There is also some evidence of significant changes in the flight periods of some species (see Table 3). However, these apparent changes are not easy to interpret for several reasons. First, and most obviously, dragonflies may have been actually on the wing both earlier and later than the dates on which they have been observed. Since recording for the two surveys was not standardised, comparisons should be seen as indica-

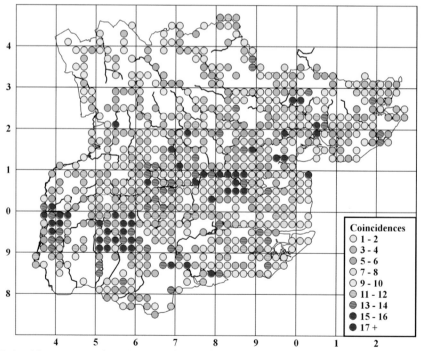

Coincidences
○ 1 - 2
○ 3 - 4
○ 5 - 6
○ 7 - 8
○ 9 - 10
◐ 11 - 12
◕ 13 - 14
● 15 - 16
● 17 +

Coincidence map showing numbers of species recorded 2000–2006, by tetrad

Species	Earliest date	Latest date
C. virgo	11/05	25/07
C. splendens	9/05	23/08
L. sponsa	24/06	15/09
L. dryas	7/06	24/08
P. pennipes	31/05	5/08
P. nymphula	17/04	11/08
I. elegans	7/05	26/09
C. puella	7/05	16/08
E. cyathigerum	7/05	30/09
E. najas	24/05	4/09
E. viridulum	7/06	4/10
B. pratense	20/05	24/06
A. cyanea	19/06	14/10
A. grandis	29/06	28/09
A. mixta	26/07	13/11
A. imperator	28/05	22/09
C. aenea	14/05	25/07
L. depressa	11/05	11/09
L. fulva	29/05	12/07
L. quadrimaculata	7/05	15/08
O. cancellatum	24/05	21/09
S. sanguineum	15/06	11/11
S. striolatum	23/06	16/11

Table 3 Flight periods of Essex dragonflies: 2000–2006

tive only. This problem particularly affects the more localised species, where infrequency of visits by observers means that both early and late dates are more likely to be missed. Another problem for interpretation is that there is a high degree of seasonal variation in flight periods from year to year. In some cases, the variation may be as great, or greater, within the periods of the two surveys than between them. Finally, occasional individuals emerge much earlier, or survive much later, than is typical for their species: first and last dates do not always reflect the flight period of the main population (generally exaggerating it).

With these provisos, however, the apparent changes are quite suggestive. Many Spring-flying species appear to emerge earlier than they did 20 years ago. These include *C. virgo, C. splendens, P. nymphula, C. puella, E. cyathigerum, C. aenea, L. depressa* and *L. quadrimaculata*. In most of these cases earliest dates are some two weeks earlier than in the previous survey. Very striking is the case of *P. nymphula*. This was always the earliest damselfly to appear in Spring, but now regularly emerges soon after the middle of April – some three to four weeks earlier than observed in the previous survey. Other early species, such as *P. pennipes, I. elegans, E. najas, B. pratense* and *O. cancellatum*, show no significant change. *L. sponsa* and *L. dryas* emerge in June, and also show no clear change in flight period.

Of the large hawkers, only *A. cyanea* is recorded as flying significantly earlier than before, but *A. imperator* now has a notably longer flight period than was reported in the previous survey. It can be seen from late May through to the last week in September – a flight period of some four months. There is evidence that *A. cyanea*, too, now continues to fly later – to mid-October – and although the latest recorded dates for *A. grandis* have not changed significantly, our impression is that more individuals survive to fly until late September than previously. Comparison of the two resident darter species is of some interest. In both surveys, *S. sanguineum* begins to fly earlier than *S. striolatum*, and finishes its flight period earlier. However, in the current survey, both species were reported earlier than in the previous survey (*striolatum* as much as three weeks earlier), while *sanguineum* continued to fly much later than before – until early November, as compared with mid-September in the 1980s.

Chapter 2
Where to see dragonflies in Essex

Dragonflies and damselflies can be found close to almost any water-body in the county: from small garden ponds, to farm reservoirs, sand-and-gravel pits, ditches, streams and rivers. Often, too, they can be seen well away from water, as they hunt for their prey in rough grassland, woodland rides and glades, parks and gardens.

However, some localities support a much richer assemblage of dragonfly species than others. The following 'tour' of the county takes you to some of the richest or most interesting sites so far studied. In some cases we have selected sites because of the large number of species that has been recorded there (usually 15 or more species), but we have also included some sites with far fewer species than this, either because the 'mix' of species is especially interesting or because the site itself is an example of a distinctive dragonfly habitat in the county. Where we include a site because it has an 'interesting mix' of species, this is generally because it has one or more 'key' species. These are species that we consider to have a specific conservation importance: possibly because they are very scarce in Essex (but may be common elsewhere – e.g. *C. virgo*), or because the Essex populations are of importance in the national context (e.g. *L. dryas, C. aenea*). Other localised species may be included in this category, even when, as seems to be happening in several cases, they are currently becoming more firmly established in the county (*L. fulva* and *B. pratense* are examples).

It is important to keep in mind that we have made no attempt to even out our survey effort across the county, with the result that some areas are much more thoroughly studied than others. You may well find sites close to you that are as rich or interesting as the ones we include here.

North-west Essex (west and north from TL7010)

This is the least well-surveyed part of the county, but it seems likely that even with more thorough surveying it would remain rather poor in dragon-fly sites and species. Much of this part of Essex was once chalk downland, but has long-since been converted to intensive arable cultivation. This, to-gether with its relatively high altitude and fast-draining soils, has resulted in a landscape less well-provided with wetlands, ponds and waterways than the rest of Essex. However, there are numerous smaller garden and farm ponds, former mineral workings and other still-water habitats that support a range of the commoner species. Many of these sites are inaccessible to the general public, and so less likely to be covered in our survey.

So, the main dragonfly interest focuses on the rivers, several of which have their sources and upper reaches in this part of the county, and the complex of ancient woodland, grassland and still-water habitats within the bounda-ries of Hatfield Forest.

1. Moving-water habitat

a. The Stort navigation

South and west from Bishop's Stortford by way of Sawbridgeworth and Har-low to Roydon has rich and varied dragonfly habitat, with some 11 species recorded (*C. splendens, I. elegans, C. puella, E. cyathigerum, E. najas, A. cyanea, A. grandis, A. mixta, A. imperator, L. depressa* and *S. striolatum*). The navigation is easily accessible by rail, with several stations along the route, and can be walked for the full distance along the tow-path. The area is under consider-able pressure from new housing development and the proposed expansion of Stansted airport.

b. The upper reaches of River Roding

Despite surrounding arable cultivation, some twelve species have been re-corded on or near the upper Roding, although these species are not neces-sarily breeding in the river itself, as there are ponds and flooded pits in the vicinity: *C. splendens, P. nymphula, I. elegans, C. puella, E. cyathigerum, E. najas, A. grandis, A. mixta, A. imperator, L. depressa, L. quadrimaculata* and *S. striola-*

tum. Lower reaches of the Roding are very rich in dragonflies: see under south-west Essex.

c. Upper reaches of River Chelmer and its tributaries the Ter and Stebbing Brook, the River Can and the Pant/ Blackwater

As with the Roding, the upper reaches and smaller tributaries of these rivers have rather fewer dragonfly species than the lower reaches, but are not without interest. The formerly very localised *P. pennipes*, for example, now occurs on these sections of the Ter, Can and Pant/Blackwater. Other species reported include: *C. splendens, P. nymphula, I. elegans, C. puella, E. cyathigerum, E. najas, A. cyanea, A. mixta, A. imperator, L. depressa, O. cancellatum, S. sanguineum* and *S. striolatum*. Again, we do not have evidence of breeding for all these species.

2. Still-water sites

Apart from garden ponds and some sand-and-gravel workings, there are two significant and publicly accessible still-water sites in this part of the county:

a. Hatfield Forest

This is owned by the National Trust, and includes a wetland reserve managed by the Essex Wildlife Trust. The Forest is described as unique in Britain, with all the features of a medieval forest retained (Rackham 1989). The large hawker dragonflies can be encountered hawking along the wide forest rides in summer, but the main dragonfly interest is in the large central lake (TL541198) and the adjacent long, curved former duck decoy. The water level in the lake was raised in 1980; it is over-hung by trees in parts, and along most of its margins has a rather narrow band of emergent vegetation. *E. cyathigerum, I. elegans* and *O. cancellatum* are abundant on the lake, but the former duck decoy is well-vegetated and sheltered, with more recorded breeding species. *C. puella* is abundant here (by contrast to the lake, which is favoured by *E. cyathigerum*), and this is a good site to observe *E. najas* and *P. nymphula*. Both 'chasers', *L. depressa* and *L. quadrimaculata*, breed here, as do *O. cancellatum, S. striolatum* and *S. sanguineum*. *A. imperator* hawks over the pond, and, from late July onwards *A. mixta* is in evidence. The other large hawkers, *A.cyanea*, and *A. grandis*, also occur in the Forest.

Several bus services between Dunmow, Braintree, Bishop's Stortford and
Chelmsford pass close to the northern boundary of the Forest, which can be
accessed on foot from Takeley via the Flitch Way (boundaries of OS Explorer
maps 195 and 183).

b. Stansted Airport lagoons *(TL549214)*

These large pools were excavated in the late 1980s to take run-off water from
the expansion of Stansted Airport. Despite some silting and occasional pollution, they already have an interesting dragonfly fauna. They are particularly significant for the presence of several of the rare migrant 'darter', *S.
fonscolombii* in the summer of 2006. These made use of extensive areas of
bare ground around the pools. *L. sponsa* occurs among the common spike-
rush on pond margins, and this is also a good site for the newly-established
E. viridulum. Other species recorded at the site include: *I. elegans, C. puella,
E. cyathigerum, E. najas, A. cyanea, A. grandis, A. mixta, A. imperator, L. quadrimaculata, O. cancellatum, S. sanguineum,* and *S. striolatum. C. splendens* also
occurs on the stream that runs alongside the main ponds.

Transport links as for Hatfield Forest.

Stansted Airport Lagoon *– where several red-veined darters were to be
seen in 2006*

North-east Essex (north and east of TL7010)

1. Moving-water habitats

The middle and lower reaches of three rivers flow through this part of Essex. The Stour, in the north, marks the boundary between Essex and Suffolk. The Colne flows west to east, via Great Yeldham, Halstead and Earls Colne, becoming tidal from Colchester eastwards. The Roman River runs just south of Colchester, flowing into the Colne estuary near Rowhedge.

a. River Stour east of Haverhill

There is rich dragonfly habitat over most of this length, but the river is especially interesting south and east of Sudbury through to Nayland.

As many as 15 species have been observed along the river, most of them established breeding species. Species observed include: *C. splendens, P. nymphula, C. puella, E. cyathigerum, I. elegans, E. najas, P. pennipes, A. imperator, A. grandis, A. mixta, A. cyanea, L. quadrimaculata, L. fulva, S. sanguineum* and *S. striolatum. C. splendens* is abundant in most places. The arrival and subsequent expansion of *L. fulva* has been well documented. This species is now abundant between Bures and Nayland, but can be found up-stream at least as far as Henny Street. *P. pennipes* is another formerly scarce species that has become increasingly common along much of the length of the Stour on the Essex border. This is also one of the relatively few places in Essex where *E. najas* thrives in a moving-water habitat.

Up-stream of Sudbury the river is less well-surveyed, but the river combines with a complex of flooded pits (**Glemsford Pits**), mostly used by anglers, and not easy to access, just north of Foxearth. So far twelve species have been recorded from this site (*C. splendens, I. elegans, C. puella, E. cyathigerum, E. najas, E. viridulum, A. cyanea, A. grandis, A. mixta, O. cancellatum, S. sanguineum* and *S. striolatum*).

The lower reaches of the river are well supplied with public footpaths, and good spots for observing dragonflies can be accessed on foot from Bures (by rail from Marks Tey or bus 753), Sudbury (rail from Marks Tey or bus 753), Wormingford (bus 753) or Nayland (bus 84).

b. River Colne

The Colne is less well-endowed with dragonflies than is the Stour, but not without interest. Between Castle Hedingham and Colchester the Colne offers quite varied dragonfly habitat, with some shallow, swiftly flowing stretches alternating with deeper, slower ones. The river broadens and deepens as it approaches Colchester, and has rich marginal, emergent and aquatic

vegetation where it is bordered by council-owned Cymbeline Meadows (TL9725 and 9825). The following species, most of which breed in the slower stretches of the river, can be seen: *C. splendens, P. nymphula, I. elegans, C. puella, E. cyathigerum, A. cyanea, A. grandis, A. mixta, A. imperator, O. cancellatum* and *S. striolatum.* From West Bergholt through to Colchester these are

The River Colne close to Colchester: locality for the scarce chaser

joined by *L. fulva* (a recent arrival, first detected in 2000), *B. pratense* (another recent arrival) and *E. najas.* There are also reports of *L. fulva* from further upstream as far as Earls Colne, and *E. najas* at Fordham (R.M. Rayner, P. Smith, D. Blakesley, T. Benton and others).

Cymbeline Meadows are easily reached on foot from Colchester, and higher reaches can be accessed at a number of points along the route of the 88 bus service between Colchester and Halstead.

c. Roman River

The Roman River rises to the north of Coggeshall and flows east to join the Colne estuary at Rowhedge. For most of its length it is a small stream with relatively little dragonfly interest, but from Copford in the west, through to Fingringhoe in the east it passes through the Roman River Conservation Zone (see Wake (ed.) 1983). Along much of this stretch of the river, and on a tributary, the Birch Brook, can be found both *C. splendens*, and the local rarity *C. virgo*. Harwood reported the presence of *virgo* at what must have been this site as long ago as 1903, and it has presumably bred there continuously since that time. There are occasional reports of *virgo* from other parts of the county, but this remains the only breeding site that we have been able to confirm (but see p.64). Along with these two spectacular species, the river supports populations of *P. nymphula*, *I. elegans*, *C. puella* and *S. sanguineum*. There are also several reports of *P. pennipes* from adjacent Friday Woods, but it is unclear whether this represents a breeding population in the river.

The best stretches for dragonfly-watching are bordered by nature reserves and accessible MoD land, with public footpaths. The area can be accessed from the west by taking the 75 (Colchester to Maldon) bus, and alighting at Heckfordbridge or Colchester zoo. From the east, the river can be approached by taking the 67 bus between Colchester and East Mersea, alighting where the bus crosses the river (OS Explorer map 184). Perhaps the best place to see *C. virgo* is the stretch of river running along the northern edge of the EWT's Roman River Valley reserve. The nearest access point is from the Layer de la Haye road, at King's Ford Bridge (infrequent Colchester to Malting Green bus service 50).

2. Still-water sites

The commonest and often richest of these are the lakes and ponds formed by the many working and former mineral extraction sites in this part of Essex. Among the larger bodies of still water can be included the major reservoirs, Ardleigh and Abberton, and a number of long-established ornamental lakes and related aquatic features. There remains a scattering of older farm- and village ponds, but unless deliberately restored and managed for conservation they are frequently overgrown, polluted and of little use to dragonflies. However, in compensation, there are many ornamental ponds created in parks, private gardens and nature reserves deliberately to encourage wildlife. Though generally small in size, these ponds can be surprisingly rich in breeding species, although access for surveying is often difficult, while the introduction of fish and encouragement of waterfowl can seriously diminish their value as dragonfly habitat. Finally, there are the 'borrow dykes' that run along the landward side of our flood defences by the sea and main river estuaries, and the associated 'fleets' that intersect the few remaining grazing marshes. These vary in their degree of salinity, and are subjected to varying

management practices. However, in total they sustain a distinctive and nationally important assemblage of dragonfly species.

Former sand-and-gravel workings: Several of these are among the richest
dragonfly sites in the county, and they illustrate an interesting process of
successive colonisation by the commoner species as they become increasingly well-vegetated. Many have been lost to land-fill and to restoration to
agriculture. Most that remain are used for angling or are included in nature
reserves or country parks. Some small-scale former extraction sites are in
use as farm reservoirs.

E. cyathigerum, L. depressa and *O. cancellatum* are often the first species to
colonise newly formed lakes, the last-mentioned being particularly prone
to basking on bare ground on the banks. Later typical colonists include *I. elegans* and *S. striolatum*, with the large 'hawker' dragonflies (*Aeshna* species),
S. sanguineum, E. najas and *C. puella* taking up residence as the aquatic flora
and fauna, emergent and marginal vegetation become fully established.
Shallow excavations, with denser stands of emergent vegetation often also
support *L. sponsa* and, sometimes flying together with it, *L. dryas*. These sites
have also been significant in the establishment and spread throughout the
county of the newcomer, *E. viridulum*.

The following are examples of former (and occasionally still active) mineral
extraction sites in NE Essex that have been the subject of study during the
present survey:

a. Fingringhoe Wick Nature Reserve (EWT) (TM0419 and 0420)

The reserve was purchased by the EWT in 1961, following some 40 years
of its previous existence as a working sand-and-gravel quarry (see Forsyth
2005). There is a large central lake, and numerous smaller ponds, some remnants of the former use of the site, others newly created and managed to
maintain the high diversity of aquatic habitat on the reserve. The terrestrial
habitat at the Wick, too, is just what the dragonflies need: a mosaic of open
heath and grassland, scrub and woodland. The tracks through the reserve
are flowery and insect rich, and are an additional hunting ground for the
adult dragonflies.

Over the years, the reserve has provided habitat for a remarkably stable
number of dragonfly species, though with some interesting changes. During

the 1980s (Benton 1988) 15 species were reported, and the figure today is 18.

Early colonisers include *E. cyathigerum* and *O. cancellatum*, and both can be seen at the Wick – the latter especially in evidence along the open banks of the lake. Two 'chaser' dragonflies (*L. depressa* and *L. quadrimaculata*) are in evidence from mid-May and breed in several of the ponds. The latter also breeds in the brackish Kit's Pond. In recent years *Brachytron pratense* has been breeding in an old farm pond on the northern edge of the reserve. Another 'speciality' of the reserve is the scarce emerald damselfly (*L. dryas*). This species is somewhat elusive, and may not breed on the reserve every year. The best places to look are the more densely vegetated ponds

Pine Pond: habitat for both emerald damselflies

with stands of common and lesser reedmace, sea club-rush, water plantain and soft rush. Its close relative, the common emerald (*L. sponsa*), is also to be seen on the reserve, often at the same ponds. Soon after the initial discovery of *E. viridulum* in north-east Essex, the species turned up at Kit's Pond, where eggs are laid in the tissues of the submerged pondweed *Potamogeton pectinatus*. There is also a strong breeding population on the lake.

All four of the common large 'hawker' dragonflies (*A. cyanea, A. grandis, A. mixta, A. imperator*) can be seen almost anywhere on the reserve during the Summer and on into Autumn. Pine Pond (down steep steps to the left of the entrance track) is a good place to watch several of these species egg-laying, along with both of the common 'darter' species (*S. sanguineum* and *S. striolatum*).

Depending on the time of year, other common species such as *P. nymphula, I. elegans* and *C. puella* may be seen almost anywhere on the reserve.

Kit's Pond: a good place to see the small red-eyed damselfly

Fingringhoe is not well served by public transport (infrequent 175 bus), but the frequent Colchester to Rowhedge service (66) gives the possibility of a pleasant footpath walk to Fingringhoe and on to the Wick.

b. Villa Farm Quarry and adjacent habitat, Alresford (TM0521)

This complex consists of remnants of Cockaynes and Villa Woods, together with quarry workings on former orchard and woodland and nearby fishing lakes. This is a very rich dragonfly site, surveyed by Robin Cottrill and others. So far there are records of 16 species: *L. sponsa, P. nymphula, I. elegans, C. puella, E. cyathigerum, E. najas, E. viridulum, A. cyanea, A. grandis, A. mixta, A. imperator, L. depressa, L. quadrimaculata, O. cancellatum, S. sanguineum* and *S. striolatum.*

Access to parts of the site may be difficult, but there is a public footpath entrance to part of it at TM056217. Rail services between Colchester and the coast stop at Alresford, and bus services 74 and 78 also pass through the village.

c. Holland Pits NR (EWT) *(TM203193)*

The reserve includes former gravel-workings and surrounding grassland and scrub, acquired by EWT in 1964. Some open areas of heath and meadow have been retained, but since the earlier survey there has been increased scrub and secondary woodland, with some pools shaded out with woody vegetation. Nevertheless, the reserve has considerable dragonfly interest, with 13 species recorded during our survey period: *L. sponsa, P. nymphula, I. elegans, C. puella, E. cyathigerum, A. cyanea, A. grandis, A. mixta, A. imperator, L. depressa, L. quadrimaculata, S. sanguineum* and *S. striolatum* (B. Seago, P. Smith, T. Benton, J. Dobson).

Bus services 7, 8, and 8a between Walton and Clacton pass through Great Holland, a short walk from the reserve. Alternatively there are footpath walks from Kirby Cross rail station.

d. Glemsford Pits complex *(see under River Stour, above)*

Reservoirs: There are two major reservoirs in this part of Essex, at Abberton and Ardleigh. For much of their length, the margins are unsuitable for many species of dragonflies, huge numbers of wildfowl affect water quality and access is limited. Nevertheless, they may have significant assemblages of dragonflies where conditions are suitable:

Abberton Reservoir

Much of the reservoir is inaccessible to the public, but good views can often be had from the causeways that run across it. There is a Nature Reserve (EWT) (TL963185), which has open grassy areas, a pond and scrub, at the edge of the reservoir. A privately owned pond at the western edge of the reserve had numerous *S. fonscolombii* present and breeding in 2006, and some individuals were also seen on the reservoir from Layer Breton causeway.

We have been given records of the following species from the nature reserve as well as other parts of the reservoir: *C. splendens, L. sponsa, P. nymphula, I. elegans, C. puella, E. cyathigerum, E. viridulum, B. pratense, A. cyanea, A. grandis, A. mixta, A. imperator, L. depressa, L. quadrimaculata, O. cancellatum, S. sanguineum* and *S. striolatum*. (R. Ledgerton, P. Smith, A. Kettle and others).

The reserve is not served by public transport.

Other ponds and lakes, including farm and garden ponds, ornamental lakes and ponds deliberately created for wildlife conservation: these are very numerous, and provide important breeding habitat for all but two or three of the Essex species. We include here some examples of larger complexes that include several ponds or lakes with varying characteristics.

a. Marks Hall Estate

This was once a large country estate, bequeathed to the public by its former owner, and since 1971 managed by the Thomas Phillips Price Trust. The central valley has a small stream, dammed at the upper end to form two large ornamental lakes. These are mainly hard-edged and not rich in dragonflies, although they provide suitable habitat for such species as *E. najas, E. viridulum, E. cyathigerum* and *O. cancellatum.* Below the lakes the stream is canalised, but has rich aquatic and marginal vegetation.

Pond where the hairy dragonfly can sometimes be seen

To the north of the lakes is a shallow pond with extensive stands of reedmace and introduced water soldier. The hairy dragonfly (*B. pratense*) breeds here, along with several other species. Along the access road to the interpretive centre are two ponds, one of which has a diverse marginal and emergent flora, including reedmace, branched bur-reed, yellow iris and purple loosestrife. Aquatic plants include Canadian pondweed and yellow water lily. Surrounding the central valley

Main lake, showing hard edges

are extensive woodlands with several smaller ponds. Some of these
that are being actively managed
have an increasingly interesting
dragonfly fauna. Among the species recorded on the estate are: *C.
splendens* (probably not a breeding species), *L. sponsa, P. nymphula*
(common, especially in the woodland pools), *I. elegans, C. puella, E.
cyathigerum, E. najas, E. viridulum,
B. pratense, A. cyanea, A. grandis,
A. mixta, A. imperator, L. depressa,
L. quadrimaculata, O. cancellatum,
S. sanguineum* and *S. striolatum* (J.
Cowlin,T. Benton and others).

There is a car park and visitor cen-
tre/pleasant café (staffed by vol-
unteers) at TL840251. For the able
bodied, there are fine footpath
walks from Coggeshall (regular
bus service 70 between Colchester
and Braintree).

*Woodland pool where southern
hawker, four-spotted chaser and
many other species breed*

Pond by the entrance drive

b. High Woods Country Park

This remnant of the former Kingswood Royal Forest was rescued from 'development' by Colchester Borough Council in 1979. A large lake was formed by damming the stream that ran through the central valley. This is stocked with fish for angling but still has some dragonfly interest. This is one of the few still-water sites where *C. splendens* sometimes breeds, and *E. najas* can sometimes be seen on the small patches of water lilies.

A woodland pond close to the interpretive centre is used extensively for educational purposes and has heavily trodden margins. Nevertheless, several dragonfly and damselfly species breed here, including *L. sponsa, P. nymphula, I. elegans, C. puella, E. cyathigerum, A. mixta, A. grandis, A. imperator, L. depressa, L. quadrimaculata* and *S. striolatum*. Another small pond was recently created on the extreme south-western boundary of the country park to take run-off water from a new housing estate (TL999264) (see photograph below).

This pond has been very rapidly colonised by a rich assortment of aquatic and marginal plants, including hornwort, reedmace, yellow iris, branched bur-reed, water dock, bistort, sea club-rush and purple loosestrife. It has a similarly rich assemblage of at least 15 dragonfly species, including:

L. sponsa, P. nymphula, I. elegans, C. puella, E. cyathigerum, E. najas, E. viridulum, A. cyanea, A. grandis, A. mixta, A. imperator, L. depressa, L. quadrimaculata, S. sanguineum and *S. striolatum*.

Balancing pond, High Woods, Colchester

There are several points of access to High Woods Country Park, but from the town centre or from North station a footpath leads alongside the railway line and under a bridge at TL998263, very close to the pond just described.

c. Five Lakes Golf Course *(TL9313 & 9413)*

There are several ponds within this complex, both near the hotel and alongside the B1026, near Salcott. The ponds are edged with dense stands of both greater and lesser reedmace. Many of the species can be observed from these roadside ponds. The following have been recorded so far: *L. sponsa, I. elegans, C. puella, E. cyathigerum, E. najas, E. viridulum, B. pratense, A. cyanea, A. grandis, A. mixta, A. imperator, L. depressa, L. quadrimaculata, O. cancellatum, S. sanguineum* and *S. striolatum.*

The Hedingham bus service 92 between Colchester and Tollesbury passes the site.

Roadside pond, Five Lakes Golf Club

Coastal and estuarine dykes and fleets: The Essex coast and estuaries have many miles of sea walls, usually with a 'borrow dyke' running just inland alongside. In some places the borrow dykes are linked to other ditches and 'fleets' that drain a wider hinterland. In some areas this hinterland is rough grazing land and of high conservation value. Elsewhere, former grazing marshes have been ploughed up and converted to arable.

Management of the dykes is very variable, as is the level and salinity of the water in them. Many are dry in the summer months. These dykes are the favoured habitat in Essex for the key species *Lestes dryas*. However, because of its tendency to inhabit ditches in the last phases of drying out, populations are quite mobile. Colonies are liable to disappear from one site only to reappear somewhere else.

a. Old Hall Marsh NR (RSPB)

This superb reserve has a great wealth of the characteristic wildlife of the Essex grazing marshes. Sensitive management has ensured the continued presence of several of the more specialised dragonfly species of this habitat. Along with Fingringhoe Wick, this is one of the best sites in north-east Essex to see *L. dryas*. The hairy dragonfly (*B. pratense*) also has an established population breeding in the borrow dykes, and other species include:

L. sponsa, I. elegans, C. puella, E. viridulum, A. cyanea, A. mixta, A. imperator, L. depressa, L. quadrimaculata, O. cancellatum, S. sanguineum and *S. striolatum.*

The location of the reserve also makes it a point of entry for migrating species – for example, two yellow-winged darters (*S. flaveolum*) seen in August 1995 (P. Charlton, I. Hawkins pers. corr.).

Access to the reserve by car requires a permit, but there is a public footpath along the sea wall. This is best accessed from Salcott (TL957132) but can also be reached from Tollesbury. There are (infrequent) bus services between Tollesbury, Witham and Colchester (Hedingham 91 or 92).

b. Holland Haven Country Park

This coastal grassland site is managed by Tendring District Council. There is a scrape and bird-hide. The dragonfly interest centres on the sea club rush-fringed ditch running through the north-eastern section of the CP, with adjacent grazing marsh, and a well-vegetated pond close to the car-park in the south-western part.

Ditch at Holland Haven where the hairy dragonfly can often be seen

B. pratense and *A. imperator* hawk along the length of the ditch, and also breed in it, as do *A. mixta, I. elegans* and *E. cyathigerum. S. striolatum* and *O. cancellatum* are also present. The pond was colonised early by *E. viridulum*, and *A. mixta, S. striolatum, S. sanguineum* and *L. sponsa* also breed here.

The site can be accessed from Holland-on-Sea, or on foot from Frinton along the sea defences (TM226178). The area between the sea defences and edge of the golf course that lies between Frinton and the country park is also of considerable wildlife interest.

South-west Essex (south and west from TL7010)

1. Moving-water habitats

a. River Roding

Swift-flowing and with a limited dragonfly fauna in its higher reaches, the middle and lower reaches of the Roding remain among the richest moving water sites in Essex. The river winds through mainly arable countryside, and can be approached at several points by public footpaths (e.g. east of Chipping Ongar, south of Loughton), or observed from the roads that cross it in several places (e.g. Passingford bridge, Abridge).

There is considerable dragonfly interest for the full length of the river from Chipping Ongar in the north to Woodford in the south, despite serious pollution incidents (such as the major chemical spill of 1985 described in Benton 1988). That event massively affected aquatic life in the river south of Abridge. However, re-colonisation has since taken place and we have records of the following 19 species for the period of the current survey:

C. splendens, L. sponsa, P. pennipes, P. nymphula, I. elegans, C. puella, E. cyathigerum, E. najas, E. viridulum, B. pratense, A. cyanea, A. grandis, A. mixta, A. imperator, L. depressa, L. quadrimaculata, O. cancellatum, S. sanguineum and *S. striolatum.*

(C. Jupp, J. Dobson)

b. The Lea Valley

The River Lea runs along the western boundary of Essex with Hertford-shire, and is designated a Country Park (comprising 10,000 acres) for much of its length. The combination of reservoirs and worked-out pits and ponds of many shapes and sizes with the moving-water habitat provided by the river itself makes the area a haven for dragonflies. Much of the land sur-rounding the river and lakes is insect-rich marsh or rough grassland, pro-viding good terrestrial habitat for adult dragonflies. The best-studied sec-tion of the valley is the Dragonfly Reserve at Waltham Abbey. The reserve includes stretches of the Old River Lea and Cornmill Stream, with cattle-grazed rough grassland between the waterways, and patches of decidu-ous woodland to the east. There are previous reports of both *C. virgo* and *C. pulchellum* from the reserve. However, we have been unable to confirm either species during the present survey. These species have been observed during the survey period:

C. splendens, P. pennipes, P. nymphula, I. elegans, C. puella, E. cyathigerum, E. najas, B. pratense, A. cyanea, A. grandis, A. mixta, A. imperator, L. depressa, L. quadrimaculata, O. cancellatum, S. sanguineum and *S. striolatum,*

(A. Middleton, T. Benton, J. Dobson, A. McGeeney, R. Woodward, S.R. Har-ris, J. Clark and many others)

The park can be accessed by walking from several nearby rail stations: Waltham Cross for Waltham Abbey and the dragonfly reserve, or Cheshunt or Broxbourne for access to sites up-stream.

The Cornmill Stream Dragonfly Reserve at Waltham Abbey

c. Rivers Can and Wid (tributaries to Chelmer) and Writtle College estate

The Writtle College estate comprises a variety of habitats including meadow, woodland, hedgerow, arable, the confluence of the rivers Wid and Can and an adjacent farm reservoir and moat. Since 1997 M. Heywood (with E.K. Sellers and R.G. Field) has monitored the dragonfly populations on the estate. In the first year 13 species were recorded (*C. splendens, P. nymphula, I. elegans, C. puella, E. cyathigerum, A. cyanea, A.grandis, A. mixta, A. imperator, L. depressa, O. cancellatum, S. sanguineum* and *S. striolatum*). Subsequently *P. pennipes* was recorded in small numbers on both the Cam and Wid, and a female *B. pratense* was seen in 2005. The regular monitoring has provided interesting information on the year-by-year fluctuations in populations on the estate.

2. *Still-water sites*

Lakes and ponds: Many mineral extraction sites, garden ponds and farm reservoirs provide important habitat for dragonflies in south-west Essex, but we focus here on a selection of publicly accessible 'complex' sites, usually with a range of water-bodies of varying condition, shape and size.

a. Epping Forest ponds

The numerous ponds scattered throughout Epping Forest have been the subject of continuous observation for their dragonfly fauna since early in the 19th century (**see chapter 4**). There are also areas of wet heathland which were more extensive in the past and supported several of the species associated with lowland wet heathland in the 19th century, some of them apparently surviving in the Forest until the mid-20th century.

Some of the ponds, such as Connaught Water and Baldwin's Hill Pond, were formed by damming small streams. Others, such as Lost Pond and Strawberry Hill Pond, were formed by mineral extraction. Wake Valley Pond had its origin in the construction of the road through the Forest, while Fairmead and Golding's Hill Ponds were probably created to provide drinking water for cattle and horses respectively. The Epping Forest Conservation Centre has compiled dragonfly records since the early 1970s, and several recorders provided detailed reports during the previous survey (reported in Benton 1988). During that period (possibly because it was the most frequently visited) Wake Valley Pond had the highest species total of all the Forest ponds, at

	Fairmead Bottom (TQ409965)	Chingford Plain (TQ395952)	Deershelter Plain (TQ425990)	Wake Valley Pond (TQ421987)	Gilwell Park (TQ385965)	Sewardstone area (TQ390960)	High Beech (TQ413983)	Long Running (TQ433977)
C. splendens	+			+		+		
L. sponsa	+							+
P. nymphula	+	+	+	+	+	+	+	+
I. elegans	+	+	+	+	+	+	+	+
C. puella	+	+	+	+	+	+	+	+
E. cyathigerum	+	+	+	+	+	+	+	
E. najas	+	+		+		+		
E. viridulum		+						
B. pratense		+		+				+
A. cyanea	+			+	+	+	+	
A. grandis	+	+		+	+	+	+	+
A. mixta		+		+	+	+		+
A. imperator	+	+	+	+	+	+	+	+
C. aenea	+	+	+	+		+	+	
L. depressa	+	+		+	+		+	+
L. quadrimaculata	+	+	+	+		+	+	+
O. cancellatum	+	+		+				
S. sanguineum		+		+	+			
S. striolatum	+	+	+	+	+	+	+	

Table 4 Dragonflies recorded at ponds in Epping Forest during the latest survey (with thanks to Andrew Middleton)

19 species (including 2 migrant species almost certainly not breeding there). Lower Forest Lake harboured 15 species, and several of the other ponds had 10 or more breeding species. Again, possibly because of intensity of recording effort, the Forest also accounted for many of the sightings of migratory and vagrant dragonflies – including *A. juncea, S. flaveolum* and *S. danae*.

During the present survey, the Forest dragonflies have been regularly studied by A. Middleton, the late C. Griffin, A. Samuels and also visited by numerous other recorders. Our impression is that the Forest retains its out-

Wake Valley Pond: breeding site for the downy emerald

standing assemblage of dragonflies with 19 resident species reported. The Forest 'speciality', the downy emerald (*C. aenea*) has been reported from 10 of the Forest ponds, apparently having become more widespread within the Forest since the 1980s. The hairy dragonfly (*B. pratense*) has re-established itself at several ponds, after an apparent absence of more than a half-century, and the newcomer to Britain, the small red-eyed damselfly (*E. viridulum*) has also reached the area. Our only two recent records of the keeled skimmer (*O. coerulescens*) are from the Forest, or very close by.

Epping Forest can be accessed easily from several directions. By rail from Liverpool Street station to Chingford station is convenient for the more southerly parts of the Forest. Alternatively the Central Line tube to Epping via Theydon Bois and Loughton gives several points of access on foot. There are also several bus routes.

b. Bedfords Park (TQ520925)

Formerly the landscaped park of a manor house, Bedfords Park is now an urban park managed by Havering Council. The park comprises marsh, woodland and flower-rich meadows, together with a large shallow lake and smaller ponds. Colin Jupp and others have consistently surveyed the main pond, most notably observing both the southern emperor (*A. parthenope*) and the red-veined darter (*S. fonscolombii*). The following species have been recorded in the park:

*Larger pond,
Bedfords Park*

*Small pond at
Bedfords Park:
habitat for
scarce emerald*

C. splendens, L. sponsa, L. dryas, P. nymphula, I. elegans, C. puella, E. cyathigerum, E. najas, E. viridulum, B. pratense, A. cyanea, A. grandis, A. mixta, A. imperator, A. parthenope, L. depressa, L. quadrimaculata, O. cancellatum, S. fonscolombii, S. sanguineum and *S. striolatum. S. sanguineum* and both *Lestes* species are found on a smaller, overgrown pond part-shaded by scrub/trees:

Bus service 500 between Romford and Harlow passes by the Country Park at Havering-atte-Bower.

c. Weald Country Park (TQ570940)

Formerly a medieval deer park, Weald Park suffered damage during and shortly after World War II. It was purchased as a country park by Essex County Council in 1953. It has a mixture of woodland, grassland and remnants of the old parkland, with a large lake and smaller lake. The smaller, more open, lake is best for dragonflies. Species recorded here include *P. pennipes* – presumably a 'wanderer' from the River Roding to the north:

C. splendens, L. sponsa, P. pennipes, P. nymphula, I. elegans, C. puella, E. cyathigerum, E. najas, E. viridulum, A. cyanea, A. grandis, A. mixta, A. imperator, L. depressa, L. quadrimaculata, O. cancellatum, S. sanguineum and *S. striolatum* (C. Jupp).

The park is situated some 2 miles north-west of Brentwood station and may be reached by a regular bus service from Brentwood.

d. Dagnam Park (TQ551928)

Owned and managed by Havering Council, there are several ponds in an ancient parkland setting. Species recorded include:

C. splendens, L. sponsa, P. nymphula, I. elegans, C. puella, E. cyathigerum, E. najas, E. viridulum, A. cyanea, A. grandis, A. mixta, A. imperator, L. depressa, L. quadrimaculata, O. cancellatum, S. sanguineum and *S. striolatum* (D. Sampson).

The park is situated on the north-eastern edge of Harold Hill, about 2 miles from Harold Wood railway station.

e. Belhus Woods Country Park (TQ565825)

Although the park dates to the mid-18th century when it was landscaped by Capability Brown, it is now managed by Essex County Council. The three lakes were formed by gravel extraction in the 1980s. Species recorded include:

I. elegans, E. najas, E. viridulum, E. cyathigerum, A. mixta, A. cyanea, A. grandis, A. imperator, O. cancellatum, S. sanguineum and *S. striolatum*.

f. Thorndon Country Park (TQ6090 (centre))

This popular and extensive country park is established on the basis of a medieval hunting park, later landscaped by Capability Brown. It has a fine mixture of ancient woodland, parkland, meadows and a marsh, with several ponds of various shapes and sizes. The Horse Pond at Hatch Farm (still privately owned) has a rich dragonfly fauna, as do Childerditch and Old Hall Ponds. A small pond at TQ621900 has the scarce emerald (*Lestes dryas*). Other species recorded include:

C. splendens, L. sponsa, I. elegans, C. puella, E. najas, E. viridulum, A. cyanea, A. grandis, A. imperator, O. cancellatum, S. sanguineum and *S. striolatum* (M. Wright, C. Jupp).

The park is approximately 2 miles from Brentwood station and can be accessed by way of a pleasant walk through Hart's Wood nature reserve, or by 565 bus from Brentwood.

2. Estuarine dykes

a. Rainham Marshes RSPB Reserve (TQ535800)

The improvements made to this reserve during the last five years have created ideal habitat for scarce species such as *L. dryas* and *B. pratense*. During the summer of 2006, *S. fonscolombii* was recorded at the site. Recorded since 2000 are the following species:

L. sponsa, L. dryas, P. nymphula, I. elegans. E. cyathigerum, B. pratense, A. cyanea, A. grandis, A. mixta, I. imperator, L. quadrimaculata, O. cancellatum, S. sanguineum, S. striolatum and *S. fonscolombii.*

South-east Essex (south and east of TL7010)

1. Moving-water habitat

The Chelmer/Blackwater Navigation Canal was built between 1793–97 to enable freight to reach the centre of Chelmsford via the sea lock at Heybridge Basin. The fourteen-mile canal (with twelve locks) runs through the countryside of the Chelmer Valley with water meadows at the western end and the arable landscape of Boreham, Little Baddow and Maldon to the east. Gravel extraction has taken place along much of the length of the canal providing breeding sites for *O. cancellatum* and *L. depressa* – species not usually associated with slow-moving rivers. During the current survey, 21 species of dragonfly were recorded along the Chelmer/Blackwater.

Most of these may be seen by walking eastwards from Ulting Lock (TL807082) towards Maldon. The white-legged damselfly (*P. pennipes*) is common where the vegetation inland of the towpath is well developed and the scarce chaser (*L. fulva*) has spread upstream from its original stronghold at Langford Golf Club. The hairy dragonfly (*B. pratense*) has been recorded along this stretch as well as at Boreham Lock (TL763086), and the nearby Waterhall Meadows (TL759053), with the small red-eyed damselfly (*E. viridulum*) found near the road bridge at Maldon (TL851078). This is the full list of 21 species recorded during the survey:

C. splendens, L. sponsa, P. pennipes, P. nymphula, I. elegans, C. puella, E. cyathigerum, E. najas, E. viridulum, A. grandis, B. pratense, A. cyanea, A. grandis, A. mixta, A. imperator, L. depressa, L. quadrimaculata, L. fulva, O. cancellatum, S. sanguineum and *S. striolatum* (numerous recorders including S. Wilkinson, S. & J. Torino, T. Benton, J. Dobson and others).

For the able-bodied, the navigation can be walked from Chelmsford centre through to Maldon (or vice versa), with a pleasant café stop at Paper Mill Lock. Alternatively the river can be accessed by walking from Danbury through EWT reserves (bus service 36 from Chelmsford), from the rail station at Hatfield Peverel, or from Boreham (bus service 71 between Colchester and Chelmsford).

2. Still-water habitats

Lakes and ponds

a. Chigborough Lakes (EWT) (TL877086)

The reserve consists of worked-out gravel pits, together with marshy areas, rough grassland and scrub. There are several shallow lakes and small ponds, together with sheltered rides and hedgerows where dragonflies may be observed hawking for prey. Species recorded here include:

L. sponsa, P. nymphula, I. elegans, C. puella, E. cyathigerum, E. najas, E. viridulum, A. grandis, A. mixta, A. imperator, L. depressa, L. quadrimaculata, O. cancellatum, S. sanguineum and *S. striolatum* (R. Neave, S. Wood, P. Smith and others).

This reserve can be reached by bus from Colchester and Maldon (service 75) and is located off the B1026 to the north of Maldon.

b. Maldon Wycke Reserve (TL840053)

This lake was created in 1998 as a balancing pond for the loss of a small pond on what is now a supermarket car park. With 17 species recorded, it has already developed into one of the best sites for observing dragonflies in the county:

L. sponsa, P. nymphula, I. elegans, C. puella, E. cyathigerum, E. najas, E. viridulum, B. pratense, A. cyanea, A. grandis, A. mixta, A. imperator, L. depressa, L. quadrimaculata, O. cancellatum, S. sanguineum and *S. striolatum* (R. Neave, S. Wood, J. Dobson and others).

The site is easily found by the A414 opposite Morrison's supermarket to the south west of Maldon.

Pond at Maldon Wycke Reserve (JD)

2. Estuarine dykes/fleets

a. Dykes north of Steeple (TL934045)

The coastal ditches and sea walls of the Dengie Peninsula are relatively remote and inaccessible (even by car). Although the former grazing marshes are now mostly converted to arable agriculture, ditches, track-sides and some remaining areas of rough grassland have considerable potential. *Brachytron pratense* was recorded on the peninsula after the previous survey, and has been recorded at several localities during the present one. However, survey work has been limited, and it is probably more widespread than our map suggests. This stretch of borrow dyke is included as fairly typical, and is notable for the presence of *Lestes dryas. I. elegans, E. cyathigerum, O. cancellatum, S. sanguineum* and *S. striolatum* have also been recorded.

b. Wat Tyler Country Park, Pitsea (TQ739867)

A large park managed by Basildon Council, Wat Tyler CP has many dykes ditches and ponds and supports a diverse variety of Odonata including the scarce *L. dryas* and *B. pratense*. Species recorded here include:

L. sponsa, L. dryas, P. nymphula, I. elegans, C. puella, E. cyathigerum, B. pratense, A. cyanea, A. grandis, A. mixta, A. imperator, L. quadrimaculata, O. cancellatum, S. sanguineum and *S. striolatum*.

The park can be reached on foot by a short walk from Pitsea rail station.

c. Hadleigh Castle Country Park (TQ799870)

This large park covers 120 hectares and comprises scrub, rough grassland and grazing marsh bordering the River Thames. The dykes and ponds are rich in dragonflies and it was here that *C. scitulum* was found at its sole British site between 1946–1953. During its flight period *Lestes dryas* is usually to be found on the ditches that criss-cross the grazing marsh on either side of the railway line. *Brachytron pratense* is another 'speciality' of the park. Species recorded during the current survey include:

L. sponsa, L. dryas, P. nymphula, I. elegans, C. puella, E. cyathigerum, E. najas, E. viridulum, B. pratense, A. cyanea, A. grandis, A. mixta, A. imperator, L. de-

*Hadleigh
Marshes,
viewed from
Hadleigh
Castle*

pressa, *L. quadrimaculata, O. cancellatum, S. sanguineum* and *S. striolatum* (I. Cotgrove, A. Middleton, T. Benton and many others).

There are several convenient access points to the country park. On foot it can be entered at the western tip close to Benfleet rail station. Several bus services run along the main A13 road through Hadleigh, from where the park can be approached from the north.

d. Foulness Island

Unfortunately it only proved possible to make one visit to this area due to security restrictions. Despite the lateness of our visit, we were able to find both emerald damselflies still on the wing. Undoubtedly further surveys would add to this list:

L. sponsa, L. dryas, P. nymphula, I. elegans, E. cyathigerum, A. mixta, A. imperator, S. sanguineum and *S. striolatum.*

Chapter 3
The species

This chapter provides accounts of all species recorded in the county – whether as established breeding species or as migrants or 'wanderers' – during the period of the current survey (2000–2006).

Calopteryx virgo (Linnaeus, 1758) – Beautiful Demoiselle

Confirmed as a breeding species on the Roman River only, the beautiful demoiselle is the scarcest of all the breeding species of dragonfly in Essex.

Description

The males have metallic blue-green bodies and, when fully mature, a blue or
purple suffusion over both pairs of wings. This is very dense over most of
the wing surface, but less so towards the base and tip. The females are metallic green, and their wings have a yellow-brown suffusion which is much less
dense than the colouration of the males' wings, and becomes darker with age.

Similar species

This species and the next (the banded demoiselle) are the only British damselflies with tinted wings. However, separating the two demoiselles from
each other can be more difficult. The difference in the distribution of the
wing-colouration should serve to distinguish the males of the two species:
in *splendens* the blue patch covers most of the outer half of each wing, leav-
ing the inner half (and the wing-tip) clear, but in *virgo* most of the wing is
densely tinted. The females can usually be distinguished by the colour of
their wing tints – green in *splendens*, brownish in *virgo*.

Male beautiful demoiselle

Female beautiful demoiselle

However, this difference is not always obvious, and there are slight colour changes as the individuals become mature. Another feature is a mid-dorsal stripe on the hindmost segments of the abdomen in female of both species: this is brown in *virgo*, and paler in *splendens*. In both species there are small white markings towards the wing-tips (false pterostigmata). In *splendens* the forewing mark is twice as long as that on the hindwing. The difference in size is less great in *virgo*. This character is difficult to make out in the field.

Flight period

During the previous survey, the flight period in Essex was given as the last week in May to the second week in August. During the present survey some individuals have been seen as early as the second week in May, but the last week in May is usually the peak emergence period.

Habitat and behaviour

This is a damselfly of swifter-flowing rivers and streams than C. *splendens* and prefers a sandy or gravel-bottomed watercourse. It can be found in shady, tree-lined stretches of river, as well as more open reaches bordered by meadows. Mature males perch prominently on emergent vegetation and seek to defend the area where the female will lay her eggs.

When a female enters a male's territory, he performs a 'fluttering' courtship display in which he hovers in close proximity, raising the frequency of his wing beats. If courtship is successful, copulation occurs, after which the female oviposits (with the male in close attendance) into plants stems such as bur-reeds, water-speedwells and water mint. The eggs take about fourteen days to hatch and the resulting larvae take some two years to complete their development. Interestingly, both *C. virgo* and *C. splendens* occur on the Roman River and may be observed (and compared) flying together over some three kilometres of the river-course. This suggests that either the habitat preferences of the two species are less specific than often thought, or that they occupy different small-scale niches interspersed along this stretch of the river.

A study of 'resource-partitioning' between these species at a site in Oxfordshire where they fly together revealed only very slight differences in the selection of perching places by territorial males (Brownett 1994). As in the

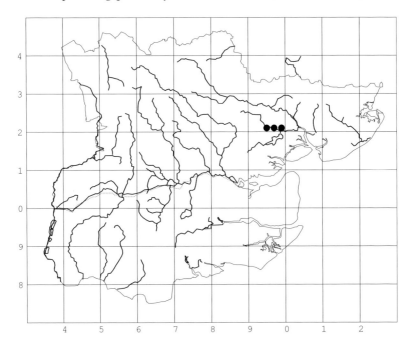

Roman River populations, there was frequent competitive interaction between males of each species both with one another and with males of the
other species.

As *virgo* appears to be more specific than *splendens* in its habitat requirements,
breeding in shallower, more swiftly flowing, gravel-bottomed stretches of
the river, it seems likely that resource partitioning between the two species
may be related more to larval than adult habitat. This was borne out in a
study of 20 breeding sites in Dorset, Hampshire and Wiltshire (Goodyear
2000). According to this study, *virgo* larvae were more frequently found in
narrower, shallower, more swiftly flowing, nitrogen-poor, oxygen-rich and
more acidic water bodies or reaches.

The study also confirmed the preference of *virgo* for stony or gravel-bottomed conditions, with little detritus or muddy deposits. There were also
differences in the mean heights of marginal and emergent vegetation, and
the amount of aquatic vegetation in the sites preferred by the two species.
However, all these differences included large areas of overlap in their toler-
ances, so that their flying together in some places is not surprising. The larvae live on or close to the river-bed, passing the winter buried in the gravel
bottom of the river. B. and J. Bailey (1985) recount an instance of terrestrial
feeding by a larva of this species. Although in this case the larva had been
taken by them from the water, the report suggests that the larvae may be
able to survive drying out of the habitat. Development is believed to take
two years.

Distribution and conservation

The beautiful demoiselle is widespread throughout Europe, except for the
extreme north and south-west. In Britain, it occurs to the south and west of a
line drawn from Liverpool to Folkestone with scattered 'outposts' in the Lake
District, North Yorkshire Moors and western Scotland. Always rare in Essex,
it is now our scarcest breeding species, recorded from just three tetrads along
the Roman River, near Colchester. Although present along the Roman River
for at least a century, the occurrence of *C. virgo* along a short stretch of just
one water system does make it vulnerable to any pollution incident that may
occur. Also, its preference for tree-lined sites with shady or dappled light
conditions means that the management of bankside vegetation should be
done sympathetically to ensure the conservation of this population, which,
along with a site near Folkestone, is the most easterly in Britain.

Early records

Doubleday (1871) reported *C. virgo* as common, 'flying in small streams', whilst Harwood (1903) recorded that it was 'not uncommon near Birch Park', undoubtedly a reference to the Roman River to the south of Colchester where it is still present 100 years later. Fitch (1879) wrote, 'on August 22nd it was quite a glorious sight to see hundreds of the beautiful demoiselle (*C. virgo*) flitting about over the Essex and Suffolk Stour, near its source' – probably near Wixoe. Mendel (1992) however, advises that this record (repeated by Lucas 1900) be treated with caution. Both Stephens (1836) and Evans (1845) classified the species now known as *C. splendens* as a variety of *C. virgo* and it is therefore likely to appear in the literature under this name. The date, during the fourth week of August, also points to *C. splendens*.

During the last century, E.E. Syms (1929) included it on his list of Epping dragonflies whilst E.B. Pinniger (1933) described it as 'not common' in the Forest. These latter reports may represent insects dispersing from colonies on the Lea and the Roding. However, Longfield (1949), whilst noting the occurrence of the species on both rivers, commented that, even in those restricted sites, the species appeared to be decreasing.

Between 1980 and 1987, the continued presence of *C. virgo* along the Roman River was confirmed by the finding of what were thought to be *C. virgo* nymphs by Adrian Chalkley in 1981, followed by the recording of adults by Ted Benton in 1982. In 1983, the presence of this species was confirmed further up-stream and also on a tributary of the Roman River. During that survey, there were three further sightings in other parts of Essex, all recorded in 1983. One was of a singleton, near Basildon (Mr Deremost, via R.G. Payne), another was of two females (observed on separate occasions) by J. Shanahan on a backwater of the Chelmer/Blackwater near Ulting and the third record was from the Old River Lea near Waltham Abbey. Here, B. Eversham observed around six individuals at a site where G.J. White also recorded them.

Please note that just as this book was going to press, we received an encouraging report of a new site for C. virgo – only the second in the county. Nine males and twelve females were observed by Mr. N.M. Rayner in a tributary to the River Colne at TL9427 and TL9428, near West Bergholt on 15th July 2006.

Calopteryx splendens (Harris, 1782) – Banded Demoiselle

This spectacular damselfly is a familiar sight on all river catchments apart from the Crouch and Roach.

Description

The body of the males is metallic blue-green, and they have a dark blue band across the central area of both pairs of wings. The wings lack pigment on the inner half and also at the tip. Females are metallic green with a green suffusion (less dense than the colouration of the male's wings) over the whole of their wings.

Similar species

Females of this species, especially, could be confused with those of *C. virgo*. See under that species for distinguishing features.

Flight period

Although the first emergence for this species in 2005 was on 28th March in Bedfordshire (Parr 2005a) and in late April in the two previous years (Parr 2004), we have no comparably early records from Essex. Peak emergence is usually in late May, as with *C. virgo*, with individuals seen as early as mid-May (the earliest recorded during the survey was 9th May). The flight period of this species seems to be longer than for *virgo*, with some individuals surviving to late August (23rd August the latest noted during the present survey).

Habitat and behaviour

The banded demoiselle is a typical species of slow-moving canals, rivers and streams – usually ones with muddy bottoms. Its preference is for open watercourses with plentiful emergent vegetation, often meandering through arable land and water meadows, but it also breeds up-stream in rivers and in small streams (see also under *C. virgo*, above). We also have several reports of this species apparently breeding at still-water sites (as in earlier literature – see below), though it is unclear whether these breeding attempts are successful. After emergence, females are often inconspicuous, resting on bankside vegetation or nearby trees and shrubs. At this time, the males occupy an area of emergent vegetation and defend it against other males.

*Male
banded
demoiselle*

*Female
banded
demoiselle*

Arriving females are courted with a distinctive butterfly-like display during which the male often lands on a potential oviposition site.

When the species is abundant (as during the current survey), males will outnumber suitable territories and large aggregations of them may be seen chasing females that have already mated or have commenced ovipositing. During copulation, a male will remove any rival sperm from a previously mated female before inseminating her with his own, thereby ensuring that he will have fertilised all the eggs laid by the female. To ensure that the female does not mate again, the male guards the egg-laying site for a period of

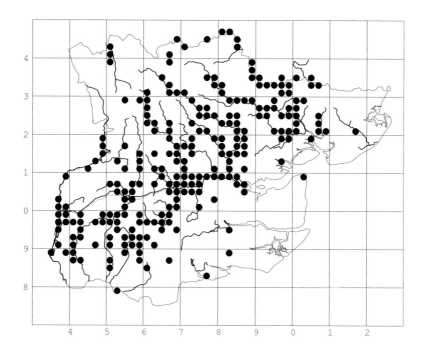

up to ten days. Eggs are inserted into a variety of plants such as bur-reeds, water-crowfoots or arrowhead and hatch after fourteen days. The larvae are stick-like and are mainly active at night. They take two years to complete their development, and over-winter buried in the muddy bottom of the river. Fully developed larvae will sometimes move some distance from the water prior to emergence. In Essex, on one occasion, three larval exuviae were found ten feet high on the trunk of an alder tree (P. Wilson, pers. comm.).

Distribution and conservation

The banded demoiselle is found in central and southern Europe, but is absent from the south-west and mountainous areas. The British population is the typical form, replaced by spp. *caprai* in most of France, central Italy and the Po plain and by spp. *balcanica* in Greece and Yugoslavia. In Britain, the main area of distribution is to the south of a line between Blackpool and Middlesbrough with isolated colonies present in parts of the Lake District and now in Scotland (Mearns & Mearns 2005).

Comparison of the results of the present survey with those of twenty years ago and earlier shows a remarkably constant distribution. The species remains common (and, in places, abundant) on reaches of slow-moving water systems away from the coast. As in the 1980s, it is still scarce in south Essex, although it was recorded from the rivers Lea, Roding and Beam, the upper reaches of the Ingrebourne and at one site on the Mar Dyke. Since the last survey, C. *splendens* has colonised the Cam in the north-west of the county and was recorded on the Holland Brook near Thorpe-le-Soken. A number of records were obtained of wandering individuals found at Bradwell, Canvey, Ashingdon, Rainham, Langdon Hills Nature Reserve and Pound Wood Nature Reserve – all at some distance from known breeding sites. If these records are included, the species was found in forty-two 10km squares, twelve more than in the previous survey.

Early records

Henry Doubleday (1871) recorded this species as 'common over small running streams' in the Epping district, whilst according to W.H. Harwood (1903) it 'abounds beside ditches and streams in all directions'. C. *splendens* appeared sporadically on Epping Forest lists subsequent to Doubleday's (Syms 1929; Pinniger 1933; Longfield 1949; and Hammond's notes). It is interesting that Longfield (1949) records that it 'breeds freely in some of the ponds in Epping Forest', and there are subsequent reports of C. *splendens* associated with still water habitat both in Epping Forest (E.P. Ryan, Wake Valley Pond, 1982 and 1983, Lower Forest Lake, 1984 (EFCC records)) and elsewhere (R. Strachan, Rivenhall lake, 1981, pers. comm.).

Elsewhere in Essex, C. *splendens* was said to be common on the rivers Lea and Roding (Pinniger 1934b, 1935 and 1937), D.A. Ashwell collected it on the Chelmer/Blackwater near Little Baddow on several dates between 1939 and 1957 (specimens in British Museum (Natural History) and Chelmsford & Essex Museum), and it was reported from the Stour at Nayland (*Essex Naturalist*, No. 28. p.152) on 26th June 1948. In 1949, Pinniger reported it at Little Parndon, presumably on the Stort (Pinniger, Syms & Ward 1950).

Lestes sponsa (Hansemann, 1823) – Emerald Damselfly

This is a fairly widespread species in Essex, but because of its rather special-ised habitat requirements it is somewhat localised. It may be significantly under-recorded.

Description

Both sexes are mainly metallic green in colour, shading to brownish in older specimens. Fully mature males have blue eyes, and pale powder-blue cover-ing the dorsal surfaces of abdominal segments 1 and 2 and 9 and 10. Females and immature (teneral) males have pale brownish eyes and lack the blue col-ouration on the abdominal segments. The pair of green markings on segment 1 of the female abdomen are roughly triangular in shape. The inferior male appendages (the inner pair, in dorsal view) are narrow and almost straight. In females, the ovipositor is short (in side view does not extend significantly beyond the tip of segment 10 of the abdomen).

Male emerald damselfly

Similar species

The scarce emerald damselfly (*Lestes dryas*) is the only other species so far recorded from Essex that could be confused with this species. The scarce emerald is usually slightly larger and more robust in appearance, but closer examination is required for definite identification especially as both species often occur at the same sites. A useful field character where fully mature males are seen is that the pale blue colouring on the dorsal surface of abdominal segment 2 covers only the anterior half or two-thirds of the segment in the scarce emerald (the whole segment in *sponsa*). The inferior male appendages in *dryas* are also broader and inwardly curved towards the tip. In females of the scarce emerald, the two green patches on abdominal segment 1 are rectangular rather than triangular, and the ovipositor is larger, projecting well beyond the tip of segment 10.

Two other species of emerald damselfly have recently been recorded in Britain, and could well be found in Essex. The southern emerald damselfly (*Lestes barbarus*) has recently been discovered in Norfolk and Kent (see **Appendix 2**). It is larger than either the emerald or scarce emerald, the males do not develop the blue colouration on the abdominal segments, and the

Female emerald damselfly

pterostigma is bicoloured (basally brown, apically whitish). The superior (outer) male appendages are yellowish with black tips, and the inferior (inner) ones shorter than those of either emerald or scarce emerald, with fine diverging projections. The willow emerald (*Lestes viridis*) has only one confirmed identification in Britain (an exuvia collected in Kent in 1992 – see J. & G. Brook 2003). As with the southern emerald, the males do not develop blue areas on the abdomen. The male inferior (inner) appendages are small, and the pterostigma is brown (black in emerald and scarce emerald, but be sure to check fully mature specimens). See Brooks (ed.) (2004) for more detail.

Flight period

According to Benton (1988) the flight period recorded during the 1980s was from late June through to mid-September. This matches the period found during the current survey when *L. sponsa* was recorded between 24th June and 15th September.

Habitat and behaviour

This species and its relative, *L. dryas*, have a distinctive resting-posture, with wings half-open, intermediate between those typical of damselflies and dragonflies. They spend much of their time at rest among emergent vegetation, or making short flights from one perch to another. The combination of their cryptic green colouration with this habit of 'skulking' among dense stands of reeds or rushes makes them difficult to spot, and they are therefore likely to be under-recorded. This habit may be a response to competitive pressure exerted by males of other damselfly species. One of us observed a male *E.cyathigerum* pick up a female *sponsa* from open water and fly with it to the edge of a pond, dropping it in water among emergent vegetation. The chastened *sponsa* flew up onto a nearby plant stem and dried its wings (3rd August 2003, High Woods, Colchester).

Characteristic habitats are shallow ponds partly or even wholly choked with emergent vegetation such as reedmace, yellow iris, common spike-rush, and, in coastal or estuarine sites, sea club-rush and glaucous bulrush. They do also occur around the margins of larger bodies of still water, such as ornamental lakes and flooded gravel-pits, but only where these have stands of emergent vegetation. The species also breeds, but less commonly, in well-vegetated slow-moving rivers and streams, such as the Cornmill Stream and lower reaches of the Roman River. The emerald damselfly is also tolerant of brackish water, and breeds commonly in coastal and estuarine ditches, where it is often found flying together with the scarce emerald and the common blue-tailed damselfly.

The females lay their eggs singly in the stems of emergent vegetation, usually in tandem with the male. Longfield (1937) reported a sequence in which the female became completely submerged during this process, but females may also oviposit 'solo', and frequently eggs are laid on plant stems well above water level. As with *L. dryas*, habitats that dry out in summer are often used.

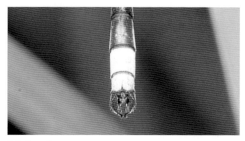

Male claspers

The eggs enter a resting phase (diapause) and do not hatch until the following April. The resulting larvae hunt for their prey (predominantly small crustaceans) among vegetation, and grow rapidly. There are typically nine larval stages (instars), and this stage lasts from two to three months. A minority of late-developing larvae do not complete their life-cycle until the following year. The adults have a long life-span, as much as two months or more (Thompson in Brooks (ed.) 2004).

Distribution and conservation

The emerald damselfly is common and widespread throughout central and northern Europe, (except Arctic Scandinavia and most of the Iberian peninsula). It is widespread and locally common throughout Britain. According to the previous Essex survey (Benton 1988), the emerald damselfly was widespread throughout the county, but rather localised, with relatively few records from the north-west of the county.

The current survey has yielded a very similar pattern (presence in thirty 10km squares in the earlier, thirty-three in the current survey). However, more intensive recording effort by Colin Jupp in TQ59 has revealed a much more dense pattern of distribution. It seems unlikely that this square is exceptional, and the species may be rather less localised elsewhere in the county than appears from our map.

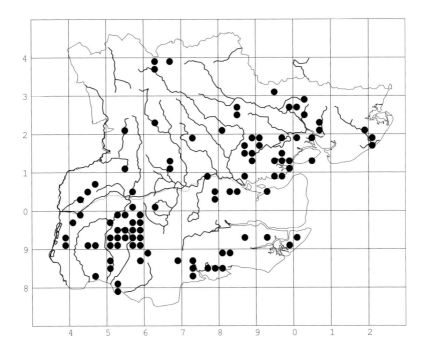

The emerald damselfly may be threatened by over-zealous management of the marginal vegetation of the ponds and ditches where it breeds, and by the in-filling of ponds. However, it is capable of breeding in summer-dry habitats, and may be expected to deal more successfully than some other species with predicted warmer and dryer conditions in future years.

Early records

Early recorders reported the presence of the emerald damselfly at Wanstead (Stephens 1835–7, Walker 1897), Coopersale Common (Campion brothers, various dates; Longfield 1949), Epping Forest (Syms 1929; Hammond, unpublished; Pinniger 1933, 1938; Longfield 1949, Pinniger, Syms & Ward 1950), Waltham Abbey (Longfield 1949), Hatfield Forest (Ashwell 1940), Warley (Pinniger, Syms & Ward 1950), Benfleet (Hammond 1937; Pinniger, Syms & Ward 1950). These mainly southern and western records are complemented by Harwood's (1903) and Longfield's (1949) reports of its presence in coastal ditches. It seems quite likely that the distribution of this species in Essex has remained stable since recording began in the early part of the 19th century.

Lestes dryas Kirby, 1890 – Scarce Emerald Damselfly

This species was considered probably extinct in Britain by 1980, but it was independently discovered at two widely separated sites in 1983, and has subsequently been recorded at other sites along the Thames estuary, coastal Essex, the south midlands and in Norfolk. However, it remains a scarce and vulnerable species, with Essex as its national stronghold.

Description

Both sexes are mainly metallic green in colour, closely resembling the emerald (*L. sponsa*). They are typically a little larger and more robust than in-

Male scarce emerald damselfly

dividuals of that species. In mature males the pale blue 'dusting' on the dorsal surface of segment 2 of the abdomen covers only the anterior half or two-thirds of the segment (contrast *L. sponsa*).

The inferior (inner) male appendages curve inwards, and are broader than those of *sponsa*. The small green spots at the front of the abdomen are rectangular in females of this species (roughly triangular in *sponsa*), and the ovipositor is larger and projects beyond the tip of the abdomen (view from the side). The eyes of the males are blue, pale brownish in females. With age, the green colouration gives way to a metallic bronze-brown.

Similar species

This species can easily be confused with *Lestes sponsa*. See under that species for distinguishing features, and separation from recently discovered rare lestids (*L. barbarus* and *L. viridis*).

Flight period

During the 1980s the flight period was reported to be from the beginning of June through to mid-August. Recording during the current survey suggests little change, with 7th June the earliest noted date, and 24th August the latest.

Habitat and behaviour

The most frequent habitats in Essex are coastal and estuarine ditches and shallow pools that are densely vegetated by sea club-rush. These are often brackish and dry out in the summer. A small minority of the Essex sites are inland ponds, sometimes in shady conditions, but also at the later stages of drying out of the aquatic habitat. After a period of apparent decline in Britain, there was considerable concern about the fate of the species by the mid-1970s. Prompted by this, N.W. Moore undertook an extensive survey. Though the species continued to occur at a few Irish localities, it was no longer found in any of the English sites (unfortunately, none of those investigated were in Essex).

Female scarce emerald damselfly

Moore attributed this catastrophic decline to a combination of three factors: loss of habitat, periods of drought, and small population numbers. In England, *L. dryas* is associated primarily (but not exclusively) with ditches and ponds on low-lying alluvial land near the sea. Sometimes it occurs in brackish water, and always in sites with plentiful emergent vegetation. Moore gives water horsetail, reedmace and, in Ireland, bulrush as plant associations for this species (Moore 1980). In Essex, *L. dryas* is more commonly found in ditches and dykes that are choked with sea club-rush. Studies along the Thames estuary by Drake (1990 and 1991) and on Old Hall Marshes (Thomas 1999) established a strong relationship between *dryas* breeding sites and ditches or pools choked with emergent vegetation, relatively saline and shallow water. In Thomas's study, larvae of *A. mixta* and *S. sanguineum*, as well as the ubiquitous *I. elegans* were found to share the *dryas* breeding sites. She also found *dryas* larvae co-existing with the small fish, the ten-spined stickleback.

Drake's studies were carried out earlier in the year (May) and he was able to observe conditions under which the larvae flourished prior to the drying out of their habitat. A canopy of stranded filamentous algae provided shade in one ditch and the larvae here were more abundant and closer to full development than in unshaded ditches. It seems that although *dryas* is strongly associated with shallow, summer-dry water-bodies, it is important for larval survival that conditions are suitable for rapid development during the spring, and that pools do not finally dry out until late May or early June. Drake mentions springtails, small water boatman larvae and water fleas as food for *dryas* larvae.

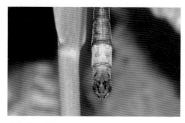

Male claspers

Thomas notes the strong correlation between sea club-rush and *dryas* breeding sites, but both she and Drake concur that it is the vegetation structure rather than the presence of sea club-rush itself that is important. At the south Essex sites, *dryas* was found in ditches where common spike-rush or common reed were abundant as emergent plants. Both studies concluded that grazing by cattle could be detrimental to *L. dryas* unless carefully controlled. Since the typical sites are transitory, representing final stages in the drying out of the aquatic habitat, the survival of the species is dependent on its ability to re-colonise new sites. If the process of drying out is accelerated by drought, with population sizes in any case low, and adjacent sites altered or destroyed by human activity, then the prospects for re-colonisation are reduced

This species and its close relative the emerald damselfly are, however, both well-adapted to summer-dry habitats, as eggs laid in the summer do not hatch until the following spring, as water-levels rise. It is thought that the scarce emerald is found predominantly in such habitats because the larvae are vulnerable to predation (Drake mentions water beetles (*Dytiscus*) and shrimps (*Gammarus duebeni*)), and few other species can tolerate these conditions. Subsequent study in Essex suggests that their ability to disperse and re-colonise suitable habitat is, fortunately, better than early pessimism suggested.

It is puzzling that in addition to its preference for relatively saline conditions in its coastal and estuarine breeding sites, *L. dryas* also breeds in freshwater woodland pools inland, as well as (especially in Essex) shallow pools in former sand-and-gravel pits. It seems very likely that its 'preference' for brackish conditions on the marshes has more to do with its ability to tolerate conditions that eliminate many competitor or predatory species there.

The adults settle with wings half-open among dense emergent vegetation, and tend to make only short, low flights between perches. It seems likely that adults in exposed marshes are vulnerable to predation by birds (Thomas mentions sedge warblers). Mating takes place in the breeding sites, and can take up to two hours. The females usually lay their eggs in tandem with their mate. She lays the eggs singly in plant stems, often beginning 30cm or more above the level of the water or mud, and works her way down, the male remaining attached to her. The following spring the larvae develop very rapidly, to enable full development and emergence before the habitat dries out.

Distribution and conservation

The European distribution of the scarce emerald is very similar to that of the emerald, except that it is much more localised in Britain. Here it is restricted to a rather small number of localities in south-east England, the south midlands and East Anglia. It also occurs in the west of Ireland.

Several sites along the lower Thames estuary in Essex were well-known to earlier generations of entomologists (see below), but since the rediscovery of this species in 1983, it has also been found in a number of inland sites, and in several places along the east coast of the county. These include Fingringhoe in the north, Mersea Island, Old Hall Marsh, several localities in the Blackwater and Crouch estuaries, the Dengie Peninsula and Foulness. In the current survey it has been found in sixteen 10km squares, compared with only

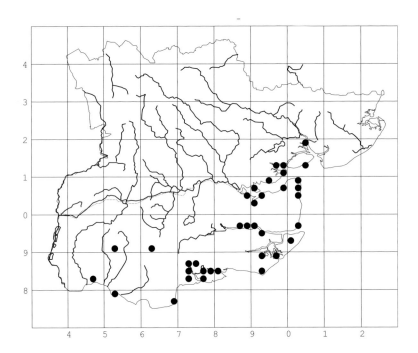

ten in the 1980s. However, this may be due to more thorough recording, and the overall pattern of distribution has remained remarkably constant.

The main threat to this rare species is intense development pressure along the Thames estuary. Climate-related drying out and inappropriate management of the coastal ditches along the east coast of the county may also be a significant threat. As the characteristic habitats are transitory, regular management of sections of the ditches where they occur to restore water levels and provide conditions for re-colonisation is needed.

History of the species in Essex

The earliest reference to *L. dryas* in the county is J.F. Stephens' (1835–7) report of it as abundant in the Thames marshes, particularly in the vicinity of Plaistow. A report by H. Doubleday (1871) that the species (under the name *Lestes nympha*) was 'rare, found on Coopersale Common' was treated with scepticism by W.J. Lucas (1900b) who commented 'one cannot help thinking there is some confusion amongst Doubleday's species of *Lestes*' (presumably recalling that Doubleday had also recorded *L. virens*, a hitherto unrecorded species, as 'rare among gravel pits'). However, both Lucas and Harwood (1903) accepted a report from Leigh in 1891 by C.A. Briggs and from Wanstead by Harcourt Bath (1892). E.B. Pinniger (1935) referred to *L. dryas* as still persisting in a 'favoured spot on the Essex marshes' and reported a visit to presumably the same locality on 12th July 1936 when *L. dryas*, 'formerly considered a great rarity' was seen 'in considerable numbers', at which time the species was also recorded near Southend and was seen in some numbers at Burnham. Hammond found the species to be 'fairly common' at Benfleet on 18th July 1937 and recorded it in his notes for that year. However, by 1949, the species was in decline and Longfield (1949) reported it as having ' a somewhat precarious hold in a few counties of England', so that the 'few isolated colonies should be most carefully preserved'.

The known sites for *L. dryas* along the Thames estuary were regularly visited during the 1950s and 1960s (the recently discovered *Coenagrion scitulum* was also present in the same area) with Benfleet being the favourite locality and it was here that Hammond last saw it in 1971 (draft letter dated 16th January 1978 in the possession of the British Entomological and Natural History Society). Other records from elsewhere in Essex were from Shenfield (1932, W.R. Fraser. BRC), South Woodham Ferrers (1950, J.H. Flint, BRC) and on the Essex/Suffolk border at Flatford Mill (1950, J.P. Dempster, BRC).

The species continued to decline until, by the mid-1970s, fears were being expressed as to its impending extinction. Out of a total of thirty 10km squares in south and east England from which *L. dryas* had been recorded, the species was found in a mere eight (two of which were in Essex) between 1961 and 1979 (Chelmick 1979). C.O. Hammond (1977) described it 'as now becoming very scarce' having 'disappeared from all its old localities near London through pollution of the habitat'. Even this review of its status seemed rather optimistic as a subsequent publication by the Nature Conservancy Council (now Natural England) declared it probably extinct (Chelmick *et al.* 1980).

Against this background of extreme pessimism, inevitably there was great excitement at the news that the species had been rediscovered by a group including Kate Rowland and Roger Payne on Foulness on 23rd June 1983. The site was on MoD land in a ditch that was almost choked with sea club-rush with other rushes, common reed and water crowfoot also present. Once the news had spread, other sites, both on Foulness and in North Kent were soon found, followed by another site (also on MoD land) to the south of Colchester by a group from the Colchester Natural History Society that included

Ted Benton. At this site, over a dozen individuals were observed and photographed as they settled low down in vegetation or made short flights in
the brackish pool and borrow dyke in which they were found. As with the
south Essex sites, the dominant plant species was sea club-rush with fennel
pondweed also present in the open area of the site (Benton & Payne 1983).

The rediscovery of *L. dryas* at several separate sites after ten years of assumed
extinction posed the question whether this influx of records was due to a
sudden immigration, or whether the species had simply been overlooked.
Accordingly, in 1984, considerable interest was focussed on surveys to confirm the presence of the species at those sites occupied in 1983, as well as
the investigation of suitable habitat elsewhere. Happily, *Lestes dryas* was still
present in considerable numbers on 27th and 29th June at the MoD site near
Colchester and subsequent visits to south-east Essex by Kate Rowland and
Roger Payne confirmed that breeding was also continuing at these sites. In
July and August of the same year, further sites were found along the Thames
estuary and Kate Rowland found a single female at Old Hall Marsh.

By the end of 1985, *L. dryas* had been recorded from ten 10km squares in south
and east Essex. The majority of sites were in coastal or estuarine marshes, in
dykes or pools commonly choked with sea club-rush. Of the sites surveyed,
several were brackish and other Odonata associated with *L. dryas* were *I. elegans* (five sites out of five), *L. sponsa* (four out of five and *S. sanguineum* (four
of five). *I. elegans* is such a widely-distributed species that the association is
not significant but the link with *L. sponsa* and *S. sanguineum* supports N.W.
Moore's findings.

Although it cannot be ruled out that *L. dryas* did become extinct in England,
with subsequent colonies becoming re-established through immigration, the
evidence seems to point against it. The species is able to colonise new sites
as old ones dry out, so its disappearance from well-established sites is not
surprising. Significantly, many of the newly discovered sites were within a
few kilometres of the previously known ones. The records from north-east
Essex (and from the Dengie and the Blackwater estuary during the current
survey) refer to populations that had previously been overlooked. As Benton and Payne (1983) concluded: '*L. dryas* is unobtrusive and easily missed
by someone not looking specifically for it.'

Platycnemis pennipes (Pallas, 1771) – White-legged Damselfly

This is a localised species of slow-flowing rivers and streams.

Description

The male white-legged damselfly is pale blue with longitudinal black markings when fully mature. There are broad, paired black markings on the final four abdominal segments. Segment 6 has a pair of black spots, and there are fine, median black lines on segments 2 to 5. The legs are white with a narrow longitudinal stripe, and the tibiae are flattened, with lateral rows of long black hairs. The eyes and antehumeral stripes are pale blue. Mature females are similar, but usually have reduced black markings, and have a creamy-white to pale greenish ground-colour. Immature adults of both sexes are pale creamy-white, and the females have much reduced black markings in this phase (f. *lactea*).

Similar species

The paler blue ground colour, and longitudinal black markings should serve to distinguish this species from all other blue damselflies that occur in Britain.

Male white-legged damselfly

Female white-legged damselfly

Flight period

In the 1980s this was reported to be from late May through to mid-August. Results from the current survey indicate little change.

Habitat and behaviour

The typical habitat for this species is the margins of slow-moving rivers and canals, with rich emergent, floating and marginal vegetation. Along suitable reaches they are sometimes very abundant and may be the most common species at some sites. However, we do have some reports of apparent breeding at still-water sites, though, as with *C. splendens*, we do not have conclusive evidence of breeding success at these localities.

Immature adults are often found in rank vegetation, or even wheat fields at some distance from the water's edge. Males seek out females, and are believed to be stimulated by their 'jerky' flight. In turn they display their white legs as a prelude to mating. The eggs are laid in the tissues of aquatic plants such as water lilies, usually by females in tandem with their mates. As in several other species of damselfly, considerable numbers may congregate to oviposit on a favoured plant. The larvae live on the river-bottom, among mud and detritus, moving up when fully developed to emerge among marginal vegetation.

Distribution and conservation

The white-legged damselfly is widespread throughout Europe except for Spain and Portugal and northern Scandinavia. In Britain it is mainly a species of southern lowland river systems, and is considered rather scarce and local. The previous (1980–87) Essex survey yielded evidence of breeding populations on only three river-systems in the county: the River Lea/Cornmill Stream, the River Roding and the Chelmer/Blackwater. Although the species seemed to be thriving on several stretches of the former two systems, it was thinly distributed and apparently declining on the Chelmer/Blackwater. However, there remained an abundant population on the Sandon Brook (passing through the Waterhall Meadows EWT reserve), a tributary of the Chelmer/Blackwater. Several searches along the Stour failed to locate a surviving population there. However, from the late 1980s onwards the species

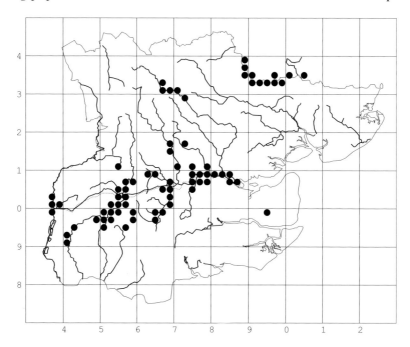

was re-found on the Stour (see Mendel 1992), and has since become abundant along that river at numerous sites between Sudbury and Dedham.

The species remains well established along the Chelmer/Blackwater navigation, and has been reported from much further up-stream on the Chelmer (as far as Hartford End), as well as on tributaries to the Chelmer/Blackwater complex (River Pant between Little Sampford and Shalford, River Wid, Roxwell Brook, River Can), the Bourne and Cripsey Brooks, as well as continuing to thrive on the Roding and River Lea/Cornmill stream. The white-legged damselfly is now known from nineteen 10km squares in the county, compared with a mere six in the earlier survey. Undoubtedly more intensive survey work accounts for some of this change, but it seems almost certain that the species has extended its range considerably within the county, and the pessimistic tone of Benton's 1988 report is no longer justified.

The species is reputed to be particularly sensitive to pollution, but Cham (2003) marshals a good deal of evidence to question this assumption. It seems likely that its requirements for lush marginal vegetation (with a suggested association with reed sweet-grass) and a suitable terrestrial 'hinterland' are more important. A study conducted on the Stour (Hofmann & Mason 2005) revealed an association between *P. pennipes* larvae and reed sweet-grass used as a perch. None were found on common reed or watercress. Despite evidence that *pennipes* can survive alongside disturbance and pollution from motorised river traffic, the rapid increase in various uses of the Chelmer/Blackwater in recent years may represent a significant threat. It will be important to continue monitoring populations of this and other species along the river.

Early records

Doubleday (1871) reported it as occurring commonly 'over streams' in Epping Forest, but it was recorded only sporadically there by subsequent observers. Apart from single records from Hatfield Forest (Ashwell 1939, unpublished) and Thorpe Hall, Thorpe-le-Soken (Oakley 1978, BRC), early records relate to four river systems in the county: the Roding (Pinniger 1935, 1937, Longfield 1949, Pinniger, Syms & Ward 1950), the River Lea/ Cornmill Stream (Lucas 1900 and Longfield 1949), the Stour (near Bures, by Roberts, cited in Lucas 1927, and Nayland, by Beaufoy from 1940 to 1960) and the Chelmer/Blackwater (Ashwell 1940 to 1970, BRC and specimens in the Natural History Museum and Colchester Museum resource centre).

Pyrrhosoma nymphula (Sulzer, 1776) – Large Red Damselfly

Often the first species to be observed in the spring, the large red damselfly is the only red species of damselfly likely to be seen in Essex.

Description

The head is black with red eyes in mature individuals. The thorax is black with a pair of red antehumeral stripes. The abdomen is red in males with black markings on segments 7–9 of the abdomen. The female has three colour forms, the commonest of which (*typica*) resembles the male but has more extensive black colouration on all abdominal segments. Of the two atypical forms, f. *fulvipes* (Stephens) has reduced black markings (although still more than the male) whereas f. *melanotum*, (Sélys) is predominantly black on the abdomen with yellow ante-humeral stripes and has been occasionally found in Essex. This latter variation is distinguished from the female red-eyed damselfly (*E. najas*) by the red joints (rather than blue) between the posterior abdominal segments.

Similar species

The small red damselfly (*Ceriagrion tenellum*) prefers sphagnum bogs and acid heathland as found in south-central and south-west England and Wales.

Male large red damselfly

Female large red damselfly

Although the small red damselfly apparently once occurred in the Epping Forest area (E. Doubleday 1835), it is unlikely that suitable habitat has been present in Essex for over a century.

Pair in tandem

Flight period

In an earlier survey, Benton (1988) recorded teneral specimens being observed on the Roman River on 21st May 1981 and suggested that 'in Essex it appears to emerge later than in other counties of southern England'. This may have been a consequence of less thorough surveying in the earlier period, but the mid-April emergence dates recorded each year in the current survey do suggest that this species (always the earliest of the spring species) is now on the wing significantly earlier than before. Nation-

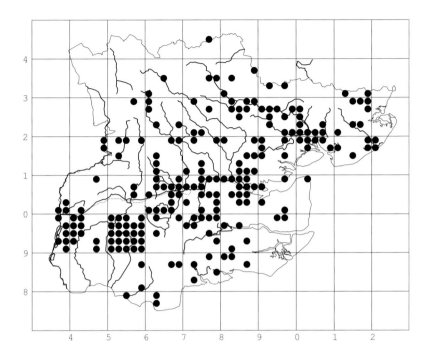

ally the first sighting of this species was on 10th April (in Hampshire, Parr 2005) in 2005, still a little earlier than our Essex sightings (17th April the earliest date during the survey). The latest date recorded in our survey was 11th August.

Habitat and behaviour

The species was recorded in a wide variety of habitats such as ponds, woodland rides, canals, rivers (Chelmer/Blackwater Navigation) and roadside ditches as well as coastal areas such as Hadleigh, Bradwell and Tollesbury. Unlike *I. elegans*, the large red damselfly appears to avoid brackish water, but it is one of the few species that breeds in the swiftly-flowing up-stream stretches of rivers and streams such as the Ter and the Roman River.

The male large red damselfly is territorial and will fly out from a prominent perch to investigate other males, egg-laying tandem pairs and females. Mating takes place during fine weather and the eggs are laid in the tissue of submerged plants, with male and female in tandem. The resulting larvae usually take two years to develop, overwintering the second time in their final stage, ready for a synchronised emergence the following spring. They live on or close to the bottom and are strongly territorial.

Distribution and conservation

The large red damselfly is widely distributed throughout Europe except for parts of the south-west and northern Scandinavia. It is generally widespread and common in most of Britain, but this is not true of its status in Essex, where it is comparatively localised and can be difficult to find, even in apparently suitable habitat. Indeed, this species has undergone a decline in eastern England during the last thirty years, probably as a result of the intensification of agriculture (Thompson in Brooks (ed.) 2004).

Despite the recording difficulties described above, the current distribution map shows this species to be more widely distributed than indicated in Benton (1988). At that time, it was recorded from twenty-four of a possible fifty-seven 10km squares, in twelve of which it was represented by a single, isolated record. Since 2000, the large red damselfly has been recorded in forty-three 10km squares, suggesting that it may have been overlooked in some areas during the previous survey. It is perhaps significant that in TQ59, which has received the greatest intensity of recording (thanks to the efforts of Colin Jupp), the species has been found to be distributed throughout the square. This suggests that this surprisingly cryptic species, with its relatively short and early flight period, maybe rather more widespread than even our current survey indicates.

Early records

The large red damselfly appeared on Doubleday's (1871) list for Epping and on F.A. Walker's (1897) list for Wanstead Park. Harwood (1903) reported it as common at Colchester, Birch, Epping 'and in many other localities'. It was later recorded from other sites in the rest of Essex: Hatfield Forest (1940, 1948, D.A. Ashwell), the River Lea (1934, E.B. Pinniger), the Stour backwater at Nayland (1968, B.T. Ward), Stanford Rivers, Margaretting, Ongar (1949, B.T. Ward) and Benfleet (1949, E.B. Pinniger).

Coenagrion puella (Linnaeus, 1758) – Azure Damselfly

This species is one of the two common 'blue-and-black' damselflies that are widespread throughout Essex.

Description

The adult male *C. puella* is predominantly blue with black markings. Typically there is a 'U'-shaped marking on segment 2 of the abdomen and blue colouration to segment 8 and the anterior half of segment 9. Typical females of *C. puella* are almost wholly black with a narrow greenish band at the anterior edge of each abdominal segment, and a black 'thistle' marking on a green background on segment 2. There is also a blue form of the female that is regularly found in Essex, in which the anterior third of each of segments 3 to 6 is blue, a feature that can cause confusion with females of *C. pulchellum.*

Similar species

In Essex the main problem is to distinguish this species from the very similar common blue damselfly (*E. cyathigerum*). The most useful feature when viewing the insects at distance through binoculars is the pattern on segments 8 and 9 of the male abdomen. Segment 8 is blue and unmarked in both spe-

Male azure damselfly

Female azure damselfly

cies. Segment 9 in *puella* is predominantly blue, but has a roughly 'w'-shaped black mark. In *E. cyathigerum* both segments are blue and unmarked. The shape of the black mark in segment 2 of the abdomen is also a useful feature if it can be seen clearly: mushroom-shaped in *cyathigerum*, 'U'-shaped in *puella*. The females are more difficult to distinguish, partly because of the variety of colour forms in both species. In side-view the thorax of *puella* can be seen to have a partial second black line, whereas *cyathigerum* has just one. The antehumeral stripes are also broader (relative to the black lines that border them) in *cyathigerum* than in *puella*. The pronotum (front portion of the thorax, looking rather like a collar in dorsal view) has a more definite mid-lobe in *puella* than in *cyathigerum*. The females of *cyathigerum* also have a small spine below segment 8 of the abdomen – absent in the *Coenagrion* species. The variable damselfly (*C. pulchellum*), sightings of which have occasionally been claimed from the Lea Valley, usually has broken antehumeral stripes (forming an 'exclamation mark') and a stalked 'wine glass' marking on abdominal segment 2, in males. The females closely resemble those of *puella* in colour pattern.

Flight period

The flight period recorded during the current survey was between mid-May and August with an earliest record on 7th May and a late record on 16th August. This is a shorter period than that for *E. cyathigerum* and *I. elegans*, the other two common 'blue-and-black' damselflies, and may have led to some small populations being overlooked – particularly later in the season when *E. cyathigerum* is abundant. During the previous survey the species was not seen before late May, compared with mid-May as a typical emergence period in the current survey.

Habitat and behaviour

The azure damselfly is found in a wide range of water bodies such as lakes, ponds, flooded gravel pits, coastal ditches and slow-moving rivers. It favours the sheltered, well-vegetated margins of such sites and also thrives in small ponds, rich in emergent vegetation but without much open water.

The males (which are not territorial) patrol prospective ovipositing sites by flying close to the water, and also watch for females from a perch in waterside vegetation. Mating takes place in the tandem position and eggs are then laid into floating vegetation. At this time, the female is vulnerable to predation from fish, newts and dragonfly larvae, and the male to airborne

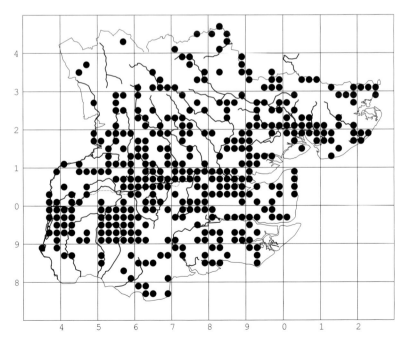

predators. Females can lay up to 4,000 eggs, which, dependent upon the temperature, hatch between two to five weeks. The larvae shelter amongst aquatic vegetation or on the bottom of the pond, lying in wait for mayfly or midge larvae and small crustaceans. The larvae usually complete their development in one year, with a synchronised emergence in May and early June the following spring (Thompson in Brooks (ed.) 2004).

Distribution and status

The azure damselfly is one of the most abundant species in Europe, where it is widespread except for parts of the south-west and northern Scandinavia. It is found throughout most of Britain, although absent from the north of Scotland. The current survey recorded the species from fifty-three 10km squares (compared with forty-three in the previous survey) with very few sites in the north-west of the county.

Early records

The status of *C. puella* has probably not changed since Doubleday's period of recording. He (1871) described it as 'common and generally distributed' in the Epping area and Harwood (1903) found it to be 'very common everywhere'. It was recorded annually between 1902 and 1909 by the Campion brothers, who also observed *C. pulchellum*-like variations (Campion & Campion 1906c), later describing these in an article in the *Entomologist* (Campion & Campion 1910). The Campions also recorded a male *C. puella* preying on a micromoth, *Tortrix viridana* (Campion & Campion 1909), and studied the larval water mites found on the nymphs of *C. puella* and other damselflies (Campion & Campion 1909). *C. puella* was subsequently listed for many other Essex sites during the first half of the 20th century – a widespread distribution that still applies today.

Enallagma cyathigerum (Charpentier, 1840) – Common Blue Damselfly

This is one of the most widely distributed damselflies in the British Isles, and at many sites the most abundant species.

Description

The male common blue damselfly has a blue abdomen with black markings. Segments 8 and 9 are both clear blue, without black markings. Segment 2 has a distinctive black 'mushroom-shaped' mark. Females can occur in both dull-green and blue forms, with arrow-shaped black markings on abdominal segments 3 to 9. There is a small spine on the underside of segment 8.

Both sexes have a pair of broad (wider than the black line that borders them on each side) antehumeral stripes on the thorax. In side view the thorax is unmarked except for the blue antehumeral stripe and thin black humeral stripe bordering it.

Similar species

In Essex, *E. cyathigerum* is most likely to be confused with *C. puella,* and both species may be found together at many sites in the county. See under *C. puella* for detail on distinguishing features.

Male common blue damselfly

Female common
blue damselfly

Flight period

E. cyathigerum has a similar flight period to *I. elegans* with the species being recorded between 7th May and the end of September during the current survey. Numbers peak during mid-summer but teneral individuals may still be seen in August and September. As with several other species, it seems that *E. cyathigerum* now emerges earlier in the Spring (by some two to three weeks) than it did during the 1980s.

Habitat and behaviour

During the current survey, the common blue damselfly was recorded from a wide range of habitats including garden ponds, ornamental lakes, reservoirs, flooded gravel workings and slow-moving rivers. It is an early coloniser of newly created sites such as gravel workings and farm reservoirs where it may occur in large numbers. Unlike *C. puella*, *E. cyathigerum* appears to prefer water bodies with large areas of open water and, on warm days, large numbers of males may be seen flying low over the water surface, often settling on patches of floating plant debris or on broken plant stems protruding above the water surface. At this time, females are more commonly found amongst marginal vegetation or at some distance from the water.

In favoured areas of emergent vegetation, males are strongly territorial although where the species is abundant, individual territories may be as little

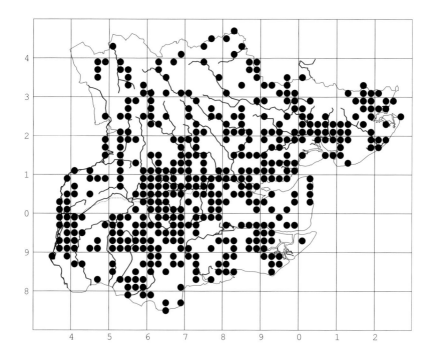

as a metre apart. Intruding males and tandem pairs are seen off (often by several males from adjacent territories) and this aggression will also extend to other species such as the red-eyed damselflies (*E. najas* and *E. viridulum*) as well as *L. sponsa*. Copulation often takes place away from the water and usually lasts around 20–30 minutes. After sperm transfer, the pair fly in tandem to seek an oviposition site which may either be close to the water's edge or over open water at some distance from it.

Egg-laying usually takes place in tandem into plant stems at the surface, or, where the population density is high, the female submerges (sometimes for over an hour) while ovipositing into the stems of underwater vegetation. Commonly, the male releases the female as she submerges, but where competition from rival males is high, he may remain attached while both become submerged (Cham 2002). When the females submerge to oviposit 'solo', they release their grip on submerged vegetation at the end of a bout of egg-laying and float to the surface. Males look out for floating females and attempt to lift them from the water. If the female is able to cooperate this is usually successful, and the male flies off with her in tandem to marginal vegetation where mating takes place (Miller 1990).

Miller notes that males frequently attempt to grasp other floating objects, and it seems likely that our observation of a male *cyathigerum* lifting a female *L. sponsa* and carrying her to the edge of a pond (see under *L. sponsa*, above) could be a modification of this 'rescue' behaviour-pattern. The eggs hatch after a few weeks and the larvae live among aquatic vegetation, lying in wait for their prey. Under normal circumstances in southern Britain development takes approximately one year.

Brownett (1990) reports the unusual occurrence of predation on this species by the grey wagtail.

Distribution and conservation

E. cyathigerum is widely distributed throughout Europe, temperate Asia and boreal America. In the British Isles, it is the most widely distributed Odonata species and, at many sites, the most numerous. The previous survey recorded this species from fifty-four 10km squares and the current survey found it in fifty-six (all but one of the 10km squares in the county). It is probable that the status and distribution of this damselfly has remained unchanged since at least the middle of the 19th century. In recent decades it has benefited greatly from the proliferation of mineral extraction sites and farm reservoirs

Early records

Enallagma cyathigerum appeared consistently on lists for Epping Forest from the time of Henry Doubleday (1871) and was considered by Pinniger (1933) to be 'probably the commonest of the small species'. Longfield (1949) included the Walthamstow Reservoirs as well as the Forest and there are numerous further records from other parts of the county: Hatfield Forest (1939, 1940, 1948, D.A. Ashwell), Benfleet (date unknown, C.O. Hammond 1936, 1949, E.B. Pinniger), River Lea (1943, C.O. Hammond); Stanford Rivers, Margaretting and Shelley (1949, B.T. Ward), and Wanstead Park (1897, F.A. Walker).

Ischnura elegans (Vander Linden, 1820) – Blue-tailed Damselfly

The blue-tailed damselfly is one of the commonest of all the Odonata in the county, inhabiting almost all but the most polluted aquatic habitats.

Description

Mature male blue-tailed damselflies have a mainly black abdomen, with a clear blue segment 8. The eyes are blue and there is a pair of blue ante-humeral stripes on the thorax. The typical form of the female is very similar, but there are several other colour forms, all of which are commonly seen in Essex. Newly emerged females have either a reddish (f. *rufescens*) or violet (f. *violacea*) colouration on the thorax. Form *rufescens-obsoleta* is the mature form of *rufescens*, with a rusty brown thorax and brown replacing blue on the eighth abdominal segment. To add to the complexity, the thorax is green in immature males, becoming blue with maturity. The adults of both sexes have bicoloured, diamond-shaped pterostigmata on the fore-wings.

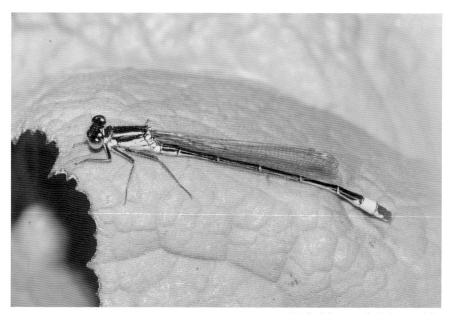

Male blue-tailed damselfly

Female blue-tailed damselfly

Similar species

Confusion may occur with several other species. Although the scarce blue-tailed damselfly (*I. pumilio*) was included on Doubleday's 1871 list, there are no other records of it occurring in Essex. However, the discovery of small populations in old chalk pits in Bedfordshire suggests that the possibility of its occurrence in Essex should not be entirely ruled out. *Pumilio* is a smaller species than *elegans*, has blue on only part of segment 8 and is wholly blue on segment 9. Additionally, the pterostigmata are shorter than in *elegans*. In Essex, it is more likely that there could be confusion with the males of both species of red-eyed damselflies, which are also predominantly black with a blue tip to the abdomen. However, the blue covers the dorsal surface of segments 9 and 10 in the red-eyed species, compared with segment 8 only in *elegans* (i.e. there is a final black segment beyond the blue in this species). The male red-eyed damselfly (*Erythromma najas*) is a larger, more robust species and in mature specimens the reddish brown eyes are always apparent. The male small red-eyed damselfly (*E. viridulum*) also has red eyes, is similar in size to the blue-tailed damselfly but has blue abdominal segments 1, 9 and 10, with blue also on the sides of segment 8.

Flight period

This species has, like *E. cyathigerum*, a prolonged flight period, and may be observed from the second week of May until the end of September. It is active even in dull, overcast conditions when other species are inactive. The flight period as recorded during the current survey differs very little from that observed during the 1980s.

Habitat and behaviour

The blue-tailed damselfly is found in many habitats such as ponds, gravel pits, reservoirs, canals and ditches, including well-vegetated, brackish borrow dykes along the Essex coast. It is most commonly seen perched or flying low down in vegetation, often some distance from water. It does not appear to be as territorial as other species, such as *E. cyathigerum*, and is often displaced from floating vegetation by more aggressive species.

Large numbers of individuals can often be seen basking on broad-leaved plants such as water lilies and burdock. Males become sexually mature

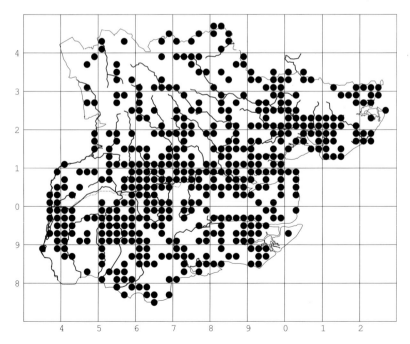

within 3–4 days of emergence – faster than for many other species – with copulation occurring amongst marginal vegetation. The females often lay their eggs later in the day when they are less likely to be harassed by males and when the reduced light, together with their dull colouration, makes them less vulnerable to predators. Egg-laying is achieved by the female (alone, rather than in tandem as with some other species) alighting on emergent or floating vegetation, before curving her abdomen under the surface of the water to lay the eggs in the stems and leaf tissue of water plants.

The resulting larvae perch among the leaves of submerged plants, and usually complete their development in one year. The adults prey on small flying invertebrates such as gnats and are themselves eaten by a variety of predators. The previous survey describes how a blue-tailed damselfly was caught in flight by a wasp, *Vespula rufa*, which consumed the head and thorax, having first severed the abdomen (Benton 1988). The unfortunate damselfly was held in position in a 'cage' formed by the legs, thorax and up-curved abdomen of the wasp. In the coastal marshes, examination of the cobwebs, notably of *Araneus* sp., shows that this species is commonly preyed upon by spiders.

Distribution and conservation

The blue-tailed damselfly is one of the commonest species to be found throughout the British Isles, extending to the north coast of Scotland. It is also found throughout central Europe, but is absent from much of Scandinavia and parts of Spain.

The earlier survey (Benton 1988) found the species in 'virtually all aquatic habitats in the county with the exception of the most seriously polluted, and some of the more swiftly flowing reaches of calcareous streams and rivers in the north-west'. Today, the status of this species remains unchanged, with records from fifty-six of the fifty-seven 10km squares across the county.

Early records

Previous references to this species suggest that it has always been as widespread and abundant as it is today. Henry Doubleday (1871) described it as 'common everywhere' and Harwood (1903) reported it as 'very common in the county, especially beside marsh ditches near the coast'.

Erythromma najas (Hansemann, 1893) – Red-eyed Damselfly

This is a rather localised species, often seen at rest on floating leaves of water plants.

Description

The mature male red-eyed damselfly is mainly black, with deep red eyes and segments 1 and 9 and 10 of the abdomen blue. There are no antehumeral stripes, and the sides of the thorax are blue. The females are much less distinctive, with red-brown rather than red eyes, and lacking the blue abdominal segments. They are predominantly black, with incomplete pale antehumeral stripes (sometimes forming an exclamation mark – as in *C. pulchellum*).

Similar species

The newly-arrived small red-eyed damselfly (*E. viridulum*) is very similar, and often occurs in the same habitats. As the name implies, this species is generally smaller and more delicately built than the red-eyed, but this is not a reliable characteristic. A useful distinguishing feature is the side view of abdominal segment 8. In males of *najas* this is black, contrasting with the blue of segments 9 and 10. In *viridulum* the blue of the final two segments tapers forwards along the side of segment 8. The side of segment 2 is also

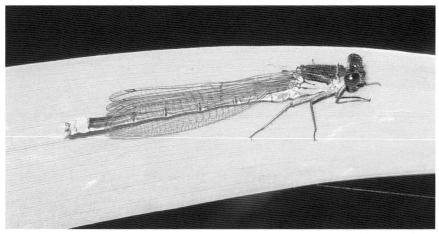

Male red-eyed damselfly

blue in *viridulum*. If a close view of the dorsal surface of the final abdominal segment is possible, this is distinctive, too: wholly blue in males of the red-eyed, blue with a black 'x'-shaped mark in the small red-eyed. The females are more difficult to distinguish, but the complete antehumeral stripes in *viridulum* (shortened or broken to form a '!' marking in *najas*) are a useful character.

The blue-tailed damselfly (*Ischnura elegans*) commonly flies at the same sites as the red-eyed, and is quite similar (see under *I. elegans*).

Flight period

In the 1980s the flight period was reported to last from the beginning of June through to early August. During the current survey, the earliest record was 24th May, with a latest date of 4th September.

Habitat and behaviour

The characteristic habitats of this species are ponds and lakes with extensive areas of floating-leaved water plants, such as water lilies, broad-leaved pond-weed and amphibious bistort, which the males use as territorial perches. In some sites, the males use floating debris or mats of algae in the absence of their more usual perches. Slow-moving rivers, canals and streams are also colonised by this species, particularly where floating-leaved plants are present. Where the red-eyed damselfly coexists with populations of the common blue damselfly (*E. cyathigerum*) there is intense competition for territory. In encoun-

ters between the two species, males of *najas* are generally defeated, but will often return to their chosen site later. However, the male *najas* and ovipositing pairs are frequently relegated to perches well away from the water margins, where competition is less intense.

Female red-eyed damselfly

Mating is prolonged, and the females generally lay their eggs while in tandem with the male. The eggs are laid in the tissue of the underside of floating leaves, or in submerged vegetation, with both insects often submerged for substantial periods of time (30 minutes has been recorded – Cham in Brooks (ed.) 2004). Ovipositing in tandem is believed to reduce the chances of the attached male being displaced by rivals, but submergence appears to carry risks of underwater predation by fish. Cham (2004c) reports attacks, at least one of them successful, on ovipositing females of *E. najas* by small rudd.

Clustering together of ovipositing pairs is believed to offer protection from predators such as frogs, and is observed in both this and other damselfly species (see Corbet 1999). Larval development is relatively slow, taking up to two years. A study by Hofmann and Mason (2005) of perching-preferences of several species of damselfly larvae on the Stour found *najas* on common reed and reed sweet-grass. In experimental conditions presented with a 'choice' of several perching sites, *najas* larvae preferred common reed stems to reed sweet-grass, water moss and water-cress.

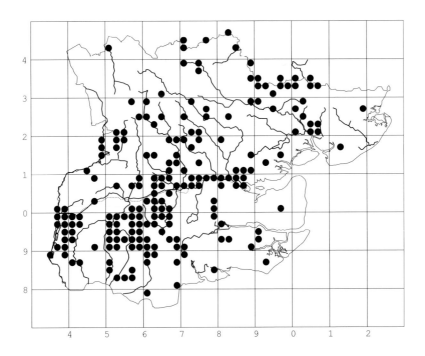

However, with the introduction of a predator (three-spined stickleback) there was a proportional shift of perches in favour of water moss and water-cress. These plants have more complex structure and so presumably offer better cover. The larvae take up to two years to complete their development. On emergence the immature adults tend to retreat from the waterside, and can often be found in nearby rank vegetation. Adults (possibly teneral) seem especially vulnerable to predation by the spider *Tetragnatha extensa*.

Distribution and conservation

The European distribution of the red-eyed damselfly is mainly central and northern (but it is absent from arctic Scandinavia). In Britain it is mainly a species of southern and midland counties of England and north Wales. The earlier Essex survey (Benton 1988) reported the red-eyed damselfly as present in only thirteen 10km squares in the county. The pattern of distribution then appeared to be mainly western, with a small number of locations in the north and north-east of the county. The current survey reveals a much more healthy picture, with records from forty-three 10km squares widely spread over the county. It is quite likely that this reflects more thorough surveying, but there is good evidence (e.g. from regularly visited sites) that this species has become more common and widespread in recent years. Our impression is that much of the recent increase has been on moving water habitats: re-establishing itself on the Chelmer/Blackwater and extending its range along the Stour, for example. Although recording effort prior to 1980 was much more limited and localised, we have the impression that the species was beginning a long-term process of spread in the county by the beginning of the 1980s and that this has continued since.

Early records

Between Doubleday's list (1871) and the 1970s there are sporadic records of the species on the Epping Forest ponds (e.g. Campion & Campion 1908; Hammond 1923–6, unpublished; Pinniger 1933, 1934a; Longfield 1949). The red-eyed damselfly was also reported from Wanstead Park, Coopersale common, the River Lea and Cornmill Stream, and Hatfield Forest up to the late 1940s by various observers, and singletons were reported from Fingringhoe Wick and Dagnam Park (L. Forsyth and D.A. Smith respectively) in 1979.

Erythromma viridulum (Charpentier, 1840) – Small Red-eyed Damselfly

This species is a very recent colonist of Britain, first recorded in this country in north-east Essex, but subsequently spreading rapidly.

Description

Mature males are predominantly black, with red eyes, and blue colouration on the dorsal and lateral surfaces of abdominal segments 1, 2, 9 and 10. The blue colouration of segments 9 and 10 extends forwards along the sides of segment 8. There is a fine black 'x'-shaped marking on the final abdominal segment (10). Both sexes have dull, greenish antehumeral stripes on the thorax. The females lack the blue colouration on the abdominal segments, have green-brown eyes, and have complete antehumeral stripes. Both sexes have a black thistle-shaped marking on abdominal segment 2.

Similar species

This species is slightly smaller than *E. najas* and can best be distinguished by the 'wedge' of blue colouration on the sides of abdominal segment 8 in the males (black in male *najas*). *I. elegans* is also very similar, but lacks the red eye-colour and has a black final abdominal segment. See under *E. najas* for more detail.

Male small red-eyed damselfly

*Male and female
small red-eyed
damselflies
in tandem*

Flight period

This is rather later than that of *najas*, with a peak from mid-July through
August. This may coincide with the life-cycle of the hornwort (see below).
However, there are much earlier and later records of smaller numbers of
individuals. During the survey, *E. viridulum* was recorded between 7th June
and 4th October.

Pair mating

Habitat and behaviour

So far, the small red-eyed damselfly seems restricted to well-vegetated ponds and lakes with surface-patches of aquatic or floating-leaved plants. The males occupy territorial perches on floating leaves, or other surface features. In both habitat and behaviour this species closely resembles the red-eyed damselfly, but there are as yet no records from moving water habitats in Britain (though they are reputed to do so in the Netherlands).

As with *najas*, immature adults as well as tandem pairs may be found among shrubs or rank vegetation away from the margins of their breeding sites. Another similarity is that, presumably as a result of competitive pressures, the perches chosen by the males are frequently out towards the middle of the ponds or lakes where they breed. Binoculars are generally a useful aid to locating and identifying them. Recent studies in Kent and in Bedfordshire suggest a very close association between the small red-eyed damselfly and hornwort (Brook 2003; Cham 2004a, b, c). This aquatic plant is rootless, overwintering at the bottom of still water habitats, but rising to the surface in late

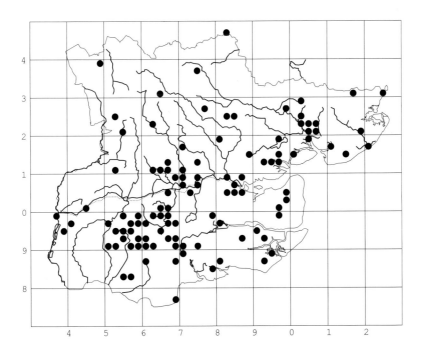

spring or early summer. Surface patches of this plant are commonly used by territorial males, and are also used for egg-laying by tandem pairs. Dredging has established that the larvae remain associated with the submerged clumps of the plant, and it seems likely that the adults emerge from exposed parts of the same plant, rather than from marginal emergent vegetation, as with many other damselflies.

Like *E. najas*, *viridulum* pairs oviposit in tandem, and frequently in groups settled together on a single patch of hornwort. However, unlike *najas*, females are not observed to become completely submerged during egg-laying. The authors observed ovipositing by this species on 25th July 2002 on the lake at Thorpe Hall, Thorpe-le-Soken. As in the above accounts, a patch of hornwort was used as the substrate by a tandem pair. Several males were also present on the same patch. The association with hornwort is presumably not obligatory as the strong breeding population at the brackish Kit's pond, Fingringhoe Wick, uses fennel pondweed for perching and ovipositing.

Distribution and conservation

The species is widespread in southern, western and central Europe, and has been extending its range in northern Europe for several decades (on the spread in the Netherlands since the early 1970s, see Ketelaar 2002). The first British record of this species was in 1999 in a pond in north-east Essex (Dewick & Gerussi 2000, reprinted with permission as **Appendix 3** to this book). In subsequent years the species spread rapidly through south-eastern England, probably aided by further immigration (e.g. a large influx in 2001 – Cham in Brooks (ed.) 2004). It has now established itself as far north and west as Bedfordshire, Berkshire and Wiltshire.

The colonisation of Essex has been very rapid, so that it is now distributed throughout the county in thirty-nine 10km squares.

Brachytron pratense (Müller, 1764) – Hairy Dragonfly

This is usually the first of our 'hawker' dragonflies to be seen in spring. Formerly extremely localised and rare in the county, it has since become more widespread.

Description

The hairy dragonfly is so-called because of the dense covering of fine hairs on the thorax and front segments of the abdomen, as well as along the ventral surface of the abdomen. The abdomen is mainly black, with pairs of tear-drop shaped markings on segments 2 to 10. These are blue in mature males and green in females. The thoracic stripes are green to blue in colour, usually complete in the males, but reduced to small yellow patches in the females. The eyes are blue in males, pale green-brown in the females and there are yellow patches on the face. The pterostigmata are very long and narrow.

Similar species

Although rather smaller, this species closely resembles other 'mottled' hawker species. The southern hawker is larger and more brightly coloured, but a more reliable feature is that in the southern hawker the blue or green patches on segments 9 and 10 of the abdomen are fused together into a continuous band (paired all the way down, or, sometimes, absent in *pratense*).

Male hairy dragonfly

The migrant hawker is very similar but usually flies considerably later in the year. It has a distinctive yellow triangular patch on abdominal segment 2, the thoracic stripes are generally reduced to small spots or dashes in both sexes, and the pterostigma is shorter and broader than in the hairy dragonfly. Despite its name the common hawker is unlikely to be seen in Essex. It is significantly larger than the hairy dragonfly and the leading edge of the wings is yellow (brown in the other hawkers).

Flight period

The species has a relatively short flight period from mid-May, through to the last week in June.

Habitat and behaviour

The breeding sites are usually unpolluted, well-vegetated still-water ponds, canals and coastal dykes and fleets as well as slow-flowing reaches of rivers. The males patrol territories, flying low along water margins, weaving in and out of patches of emergent vegetation. The patrolling behaviour of males in an Essex coastal grazing marsh was described by Benton (1988):

'Two males were observed for some time, hawking along adjacent sections of a shallow borrow dyke which was almost choked with sea club-rush. The flight was very swift and low down, just above or even below the tips of the

Female hairy dragonfly

rushes, with rapid direction-changes that deceived the eye. The dragonfly appeared much smaller and far less conspicuous than I had been led to expect from text-book illustrations. Flight was abandoned immediately full sunshine was lost. In fact, the insect seems even to anticipate the obscuring of the sun

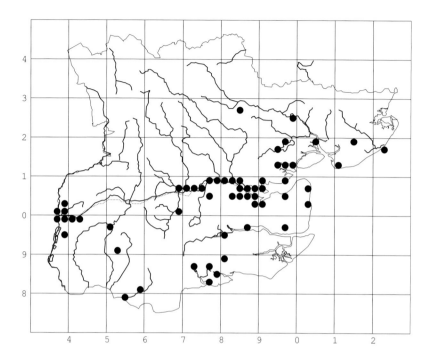

by advancing cloud, and simply disappears into dense beds of sea club-rush'.

Patrolling males clash with others of their own or other species, such as 'chasers' (*Libellula* species) or emperors (*Anax imperator*), which are also on the wing early in the season. Passing females are grasped in flight, and mating takes place with the pair settled in nearby rank vegetation. After mating the female soon commences egg-laying close to the water's edge, inserting the eggs into dead or living plant tissue. She usually oviposits 'solo', but sometimes with the male hovering in attendance. Under favourable conditions larval development can be completed in a single year (e.g. see Holmes 1984), but more usually the larval stage lasts for two or more years.

Both males and females hunt for invertebrate prey on adjacent areas of grassland. Damselflies, sawflies and mayflies are among their reported victims. One observed near Frinton, Essex, caught a worker bumblebee, but had to settle for some time to consume it (TB, field notes, 09/06/05), and we have a report (S. Wilkinson) of one eating *I. elegans*.

Distribution and conservation

The hairy dragonfly is widespread through central and eastern Europe, but absent from most of Scandinavia, and very localised in the south and west. In Britain it has always been regarded as a rather scarce species. It declined in the post-war period, but in recent years it has recovered much lost ground for reasons that are not easy to establish (though a series of warm summers from 1991 has been suggested – Perrin 1999). It is currently widely distributed but still local across large areas of East Anglia, south-eastern England, parts of the south-west and south Wales, and, much more sparsely, northwards to south-west Scotland.

During the 1980s it was the subject of intensive searches in Essex, focussing on former localities. These proved fruitless until 28th May 1985 when Ted Benton and Kate Rowland were able to visit the MoD land on Langenhoe marsh. The above-quoted passage refers to that occasion. On 18th June the same year Benton and Alan Wake observed a male at the same location and on 25th June the same team observed pairing and ovipositing.

So, by the late 1980s this species appeared to be on the verge of extinction in Essex, with only one known breeding site. However, by the mid-1990s a general recovery was underway in other parts of Britain, with new records from Hertfordshire, North Warwickshire and Nottinghamshire (BDS *Newsletter* 29, Spring 1996). This was reflected in Essex at the same time, with reports of fresh sightings from three areas in the county: the Lea Valley in the south-west, the Southend area, and several localities close to the coast in north-east Essex.

Recording work in 1997 witnessed a consolidation of this expansion of range in the county, together with the reappearance of the hairy dragonfly on its old locality along the Chelmer/Blackwater. In the north-east, there were reports from Old Hall Marsh RSPB reserve, Tollesbury Wick EWT reserve, Curry Farm nature reserve, near Bradwell, and along the Bradwell and Asheldham brooks on the Dengie Peninsula (G. Smith, P. Charlton, own obs. and others). Iris Cotgrove reported seeing territorial male hairy dragonflies on the River Chelmer on 6th and 8th June in the same year, and the late Geoff Pyman reported sightings on the Sandon Brook, a tributary to the Chelmer/ Blackwater the following year. These were the first mid-Essex sightings since Ashwell's records between 1940 and 1960.

Since then, the hairy dragonfly has continued to strengthen its hold in the above localities, and has appeared in a scattering of localities inland as well as along the coast and estuaries.

It is now found along the Chelmer/Blackwater from Chelmsford through to Maldon, at a cluster of sites in the Lea Valley and in nearby Epping Forest, and in several apparently isolated inland colonies such as at Marks Hall, Cogge-shall. Breeding is confirmed at several ponds in Epping Forest (notably Wake Valley and Little Wake Valley (A. Samuels and others)). However, its strong-holds in the county appear to be the sea defences and remaining fragments of grazing marshes along the coast and estuaries. During the current survey, the species has been recorded from twenty-one 10km squares in the county.

Early records

These indicate a similar pattern of distribution to that of today prior to World War II, with an apparent decline almost to extinction between then and the late 1980s. Epping Forest has the most continuous history of recording, and it was reported there by a succession of observers from E. Doubleday (1835) through to E.B. Pinniger's report for 1934, almost a century later ((Stephens 1835-7, H. Doubleday 1871, Campion & Campion 1906c, 1907, 1909a, 1909d, Hammond 1923-6, 1928, unpublished).

According to Pinniger it was rare in the Forest prior to 1933 but in that year appeared in 'good numbers' at virtually every pond in the forest (Pinniger 1934b, 1935). We have no data on subsequent sightings in the Forest until the 1990s, but Longfield (1949) referred to it as occurring on the 'Lea Canal'. It is possible that this was always its stronghold in the south-west of the county, with some periodic extensions into Epping Forest.

Harwood (1903) referred to the species as 'widespread but seldom met with', and gave Colchester and St. Osyth as localities. In 1975 it was reported from Langenhoe Marsh by A.C. Warne, so presumably maintained a small, local-ised population on the Essex marshes through to the present. The only other post-war reports prior to 1985 were Ashwell's observations on the Chelmer/ Blackwater, near Little Baddow from 1940 to 1960, and a 'possible' sighting by Pinniger in 1949 at Benfleet (Pinniger, Syms & Ward 1950).

Aeshna cyanea (Müller, 1764) – Southern Hawker

A large inquisitive insect, the southern hawker will readily approach and investigate an observer intruding into its feeding territory.

Description

Males have two vivid broad, yellow-green antehumeral stripes (occasionally taken for eyes by the uninitiated!) on the thorax and pairs of yellow spots on abdominal segments 3–8. The sky-blue markings on segments 9 and 10 are fused forming a band across the dorsal surface of the abdomen. The female is very similar, except that the pale markings on the abdomen are green throughout, and the eyes are green-brown (blue in the males). Both sexes have a pale triangular marking on segment 2 of the abdomen.

Similar species

Both *B. pratense* (which flies earlier) and *A. mixta* have the pale spots on the abdominal segments paired throughout (or reduced/absent on the final segments), whilst in *cyanea* they are fused into bands on segments 9 and 10. *A. cyanea* is also a noticeably larger insect than the other two species. The large,

Male southern hawker

Female southern hawker

bright 'eye' markings on the thorax are also distinctive (narrower in male *A. mixta*, much reduced in both female *mixta* and in *pratense*). The so-called common hawker (*A. juncea*), has very occasionally been seen in Essex. This is another large species, but the pale abdominal markings are paired throughout, both sexes lack the prominent pale triangular marking on abdominal segment 2, and the leading edge of the forewing is yellow (brown in *cyanea*).

Flight period

During the current survey, *Aeshna cyanea* was observed from late June to the end of September. Extreme dates reported were 19th June to 14th October. Typical emergence dates may be a little earlier than reported in the 1980s, and the normal flight period somewhat longer. However, Benton (1988) reports an exceptionally late sighting – 26th October 1986.

Habitat and behaviour

The southern hawker may commonly be seen hawking a regular beat along a woodland edge, sheltered hedgerow, mature garden or other suitable habitat where insect prey is plentiful. It often follows the contours of a hedge,

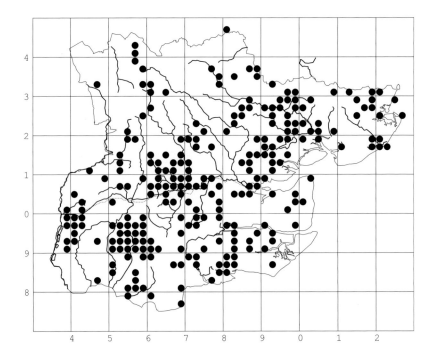

frequently descending close to ground level in pursuit of prey. Unlike some other hawkers, which fly continuously for long periods, *A. cyanea* may frequently be found (and photographed!) perched on sheltered vegetation. When hawking along a woodland ride it will often fly up to and investigate a human passer-by before flying off.

Most still or slow-moving water systems are used for breeding including woodland and garden ponds, flooded gravel pits, canals, dykes and farm reservoirs. The males are aggressive and vigorously defend their territories against intruding rivals. They tirelessly patrol 'beats' along a pond margin or river bank in search of females, chasing off other hawkers of their own and other species, as well as smaller 'darter' and 'chaser' dragonflies. They frequently disrupt the egg-laying of darters in their territories, and will attempt to mate (usually unsuccessfully) with females of their own species as they oviposit. Females visit the breeding ponds, particularly on warm afternoons (McGeeney in Brooks (ed.) 2004), where they are seized by a male, the pair flying off in the 'wheel' position. Mating, which may take up to two hours, often occurs high up in surrounding trees at some distance from the water. Ovipositing follows, with eggs laid singly into substrates such as dead plant tissue, damp mud and submerged logs. Brown substrates are reportedly preferred, although less discriminating behaviour is often encountered. Philpott (1985) reported: 'I have seen it ovipositing in some peculiar places. For example in my garden shed, in the fence and also in my ankle, through my sock while I was standing at the edge of the pond'. He goes on to say that the socks were dark-brown and Dunn (1985) described a female that 'proceeded to oviposit in the arm-folds of my jumper', repeatedly returning to the brown jumper after being released into the air. The attraction to brown substrates is called into question by this extract from the field notes of one of us: '*A. cyanea* f. ovipositing in damp mosses on the bank. It flew up to inspect me, first settling on the frame of my glasses, then my head, and next, in turn, attempting to oviposit on my green shirt, khaki field trousers, blue fleece and, finally, the skin of my arm!! Presumably ovipositing females have a very high state of generalised arousal for this activity and will test out pretty well any substrate in the vicinity of the pond'.

The eggs overwinter and hatch the following spring, with the larvae feeding on invertebrates amongst the leaf litter and dead plant material at the bottom of the pond. As the larvae grow in size they prey upon tadpoles and small fish, and are ready to emerge during their second spring. After dark on a warm evening, the mature larvae climb the stem of a reed or twig and complete their final transformation before dawn.

Distribution and conservation

The southern hawker is found throughout most of Europe, but absent from northern Scandinavia and parts of southern Europe. In Britain, it is common in the south, less frequent in the north and rare in Scotland. The previous survey found *A. cyanea* to be evenly distributed throughout the county, with records from forty-six 10km squares. The current survey reaffirmed its distribution with records from fifty-one 10km squares. It is probable that its status has remained largely unchanged at least since Doubleday's time.

Early records

Doubleday (1871) described the species as 'very common' in the Epping area and it also appeared on F.A. Walker's (1897) list for Wanstead Park. The Campion brothers recorded it annually between 1902–09 and Harwood (1903) stated that it was 'by far the commonest as well as one of the handsomest and most widely distributed of its family'.

Aeshna grandis (Linnaeus, 1758) – Brown Hawker

This widespread and distinctive dragonfly is the only brown hawker to be found in Essex.

Description

Both abdomen and thorax are brown, and the wings are slightly tinted orange-brown. There are no antehumeral stripes, but two yellow bands on each side of the thorax. Males have inconspicuous small blue patches on the sides of the abdomen and also on the dorsal surface of segment 2. The female also has small pale markings on the abdomen, but these are yellow in colour. The eyes are brownish with blue tints in the males and yellowish in the females.

Similar species

The Norfolk hawker (*Aeshna isosceles*) is the only other brown dragonfly to be found in Britain, but it is confined to the Suffolk and Norfolk broads. It lacks the orange-brown suffusion of the wings, has green eyes, and a yellow triangular marking on segment 2 of the abdomen. We know of no Essex records.

Flight period

The flight period (from the end of June to the end of September) is very similar to that for *Aeshna cyanea*. The earliest date recorded during the cur-

Male brown hawker

rent survey was 29th June with the latest report on 28th September. There appears to be little or no change since the 1980s.

Habitat and behaviour

In Essex, the brown hawker is a species that favours flooded gravel pits, lakes, ponds, and slow-moving rivers, colonising newer water bodies after the marginal vegetation has become established. It also seems to be tolerant of pollution and at some sites may be one of the few species present. However, where it occurs on coastal marshes (as at Old Hall) it seems to be confined to the fresh-water ditches.

This species is commonly observed hawking over open water – its gliding flight and shallow wingbeats combined with faster bursts of activity as it pursues its insect prey. It may also be found away from water, feeding in gardens, over rough ground near to the breeding pond or along woodland rides or hedgerows. When engaged in this activity several individuals – six or more – may share a patrol beat and continue to fly tirelessly, feeding on the wing.

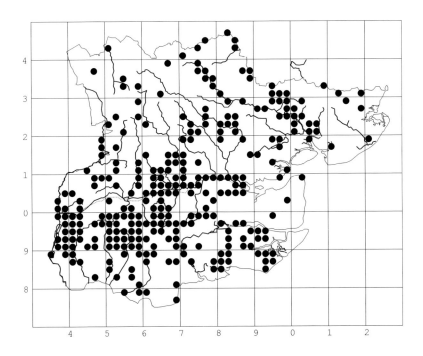

However, from mid-afternoon, or when clouds obscure the sun, they disappear – sometimes settling in rank vegetation, sometimes hanging higher up on hedgerow twigs or lower branches of trees. If warm sunshine returns, they begin to 'hawk' again after two or three minutes. However, they may also be seen flying over breeding sites in quite cool, dull weather. We have reports of one catching and eating a large white butterfly, and of another caught and eaten by a kingfisher (S. Wilkinson).

The males seldom interact aggressively, with breeding territories well-spaced out around the edges of ponds. Copulation usually takes place during the morning, after which the female lays her eggs singly, unaccompanied by the male. Typically she chooses an ovipositing site that offers support such as a floating log, emergent stem or other surface debris from which she can curve her abdomen down to the substrate into which the eggs are laid. Occasionally, group oviposition has been recorded with a number of females egg-laying into the same dead log as others either perched nearby or swarming around as though waiting their turn. As one female completed egg-laying, the vacant oviposition site was taken up by one of the attendant females (Tyrrell 2004). The eggs overwinter in diapause, before hatching the following spring. The larvae take 2–4 years to develop, depending on the water temperature and the availability of food.

Distribution and conservation

The brown hawker is widely distributed in central, eastern and northern Europe, but absent from most of Italy and the south-west. In Britain it is absent from the far south-west, Scotland and large parts of northern England. During the previous survey it was recorded from 'forty-six 10km squares in Essex, including the north-west, but it is seemingly absent or scarce, on the east coast of the county' (Benton 1988). Twenty years later, the distribution is remarkably similar, with records from fifty-one 10km squares and, again, an absence of records from the brackish coastal dykes bordering the North Sea.

Early records

Doubleday's (1871) Epping list gave it as common, it appeared on F.A. Walker's (1897) Wanstead Park list, and Harwood (1903) claimed: 'probably it is acquainted with every parish in the county'. It was recorded annually between 1902–09 by the Campions and the results of the two recent surveys support the continued validity of Harwood's comment.

Aeshna mixta Latreille, 1805 – Migrant Hawker

This late-flying hawker has become increasingly common and widespread in Essex. An earlier English-language name, 'scarce hawker', is no longer applicable.

Description

The thorax of the migrant hawker is brown, with reduced yellow ante-humeral stripes. The abdomen is brown, shading to black, with paired blue (in males) or green (females) rounded spots in segments 4 to 9 or 10. Segment 2 of the abdomen has a yellow triangular marking. In males there is also a blue band across abdominal segment 2 (paired yellow spots in the female). The eyes are blue in the male, greenish brown in the female.

Similar species

The hairy dragonfly (*B. pratense*) has a similar pattern of paired spots on the abdomen, but lacks the yellow triangle on segment 2. It also flies earlier in the year than *mixta*. The so-called common hawker (*A. juncea*) is also similar in pattern, but lacks the yellow triangle on segment 2 of the abdomen and is rarely seen in Essex. The southern hawker (*A. cyanea*) is common in Essex and has a flight period that overlaps that of the migrant hawker. It is a larger and more brightly coloured insect, and has well-marked thoracic stripes. How-ever, the pale markings on the final two abdominal segments of the southern hawker are fused to form a band, but remain separate in the migrant hawker.

Flight period

During the 1980s the flight period recorded in Essex was from the be-

Male migrant hawker

ginning of August to the end of October. In the current survey it has been seen as early as 26th July, and as late as 13th November. It seems quite likely that the flight period has been extended in both directions over the two decades. This species and the common darter (*S. striolatum*) are usually the last species on the wing in the autumn.

Habitat and behaviour

In Essex, coastal and estuarine borrow dykes were formerly the most favoured breeding sites, but it has since established itself in a wide variety of habitats. These include: flooded pits and farm reservoirs, a wide variety of ponds and lakes, ditches and slow-flowing rivers, canals and streams. Adults can often be seen away from water, patrolling stretches of hedgerow or woodland rides. In this mode they are not territorial, and large numbers often congregate to feed where prey are numerous – for example on emergence dates for winged ants. They take time out from hawking more frequently than other hawkers, and can sometimes be seen 'sun-bathing', hanging from branches

along sheltered woodland rides or open glades. This is particularly common in late afternoon on sunny days.

Males patrol stretches of water margin, flying low and often weaving in and out of emergent vegetation. Here, they do exhibit some territorial behaviour, with occasional 'interference' competition when patrolling males encoun-

Female migrant hawker

ter each other. However, this is much less vigorous than the territorial conflicts observed in male *A. cyanea* and *A. imperator*, and several males may take up territories close together on the margins of even quite small ponds. When females are encountered at the water's edge they are grasped in flight. The mating pair settle among stands of reedmace, sea club-rush, or other nearby vegetation, flying off in the 'wheel' position to find another hiding place if disturbed. Video recording by Gibson (2003 & 2006) of the behav-

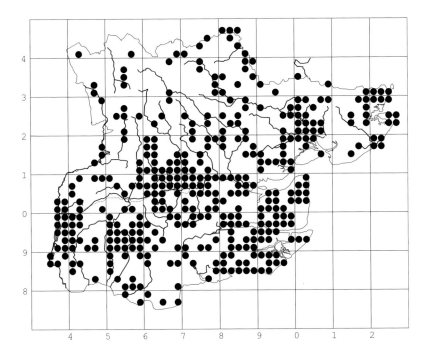

iour of pairs of *A. mixta* reveals a complex series of events. An early phase is characterised by thrusting movements of the male abdomen, presumably associated with the transfer of sperm to the female. Later, movements initiated by the female are interpreted as signals to the male that she is ready to commence egg-laying. The whole sequence was reported to take up to 20 minutes. The females lay their eggs 'solo' in plant tissue while holding on to plant stems or floating debris. In this activity they are often concealed among the stems of emergent plants in the shallow margins of the pond or ditch, and will often lay their eggs among mud and mosses on the bank, or in plant tissue above the water level. Their presence can often be detected by the rustle of their wings against plant stems.

The eggs do not hatch until the following spring, but the larvae develop rapidly and are fully grown by late July or August, when they climb up plant stems and complete their metamorphosis. Although *mixta* now breeds in a wide variety of habitats, it remains one of the few species that can tolerate brackish coastal ditches and fleets. We have several reports of larvae in coastal borrow dykes (P. & P. Wilson, A.C.J. Wood) and one of larvae in the same sampling site as those of *L. dryas* (Thomas 1999) The immature adults leave the breeding sites, and spend much of their time basking in sunshine, or hawking for prey. When basking they hang vertically from a convenient perch such as a twig or plant stem.

Distribution and conservation

The migrant hawker is widespread in southern, western and central Europe, but was probably present in Britain only as a migratory species before the end of the 19th century. It seems likely that most sightings in the early decades of the 20th century were also of migratory individuals, but there was a gradually expanding breeding population in south-eastern England. Prior to 1977 almost all records were south and east of a line from the Wash to the Bristol Channel. However, by 1990 there had been an expansion of range to Yorkshire in the north, and Pembroke in the west (BRC in Hammond 1977; Merritt, Moore & Eversham 1997). Subsequently there has been a further extension of range and in most years there are reports of (often large-scale) immigrations along the east coast (e.g. Parr 2002, 2003, 2004).

Essex seems to have been one of the earliest of the English counties to have been colonised, so it was already widespread and common as a breeding species when the 1980/87 survey took place. However, searches in the north-west of the county failed to locate the species there, giving a total of fifty-one 10km squares occupied across the rest of the county. During the current survey the migrant hawker has been found also in the north-west, with a presence in all fifty-seven 10km squares in the county.

Early records

The first reference to *A. mixta* in Essex is in J.F. Stephens (1835–7), and the locality mentioned is Epping. W.F. Evans (1845) repeated Stephens' record but it did not appear in E. Doubleday's (1835) list. H. Doubleday's (1871) record is probably erroneous. W.J. Lucas (1900a), mentioned reports from W.H. Harwood of *A. mixta* in the Colchester area, the species having been scarcer than in the previous year. Lucas (1901) commented on the comparative abundance of this 'usually scarce' species in the 1900 season, mentioning sightings at Pitsea (H.J. Turner) and Loughton (F.M.B. Carr). Harwood himself (1903) commented that the species was considered 'rare and local', but 'during the last two years has extended its range and appeared in larger numbers'. He continued: 'It is now well distributed in the Colchester district and as all the inland examples seem to be mature, while the majority of those found on the coast are immature, this may afford a clue to a fuller knowledge of its earlier stages than we at present possess'. Presumably Harwood was suggesting that the breeding sites for *A. mixta* were then mainly coastal.

The species was observed sporadically in Epping Forest by the Campion brothers (1902 and 1906), but does not appear on E.E. Syms' (1929) list. E.B. Pinniger (1933) clearly regarded the status of *A. mixta* as puzzling. In a note devoted to this species (Pinniger 1934a) he reported several searches for the species in 1933 which 'failed to produce any definite evidence'. The species did finally turn up in Pinniger's lists for 1935 ('the first record there for some years') and for 1936 (Pinniger 1936 and 1937), and was reported as having been common in the Forest in 1937 (Pinniger 1938). Interestingly, *A. mixta* first appeared in C.O. Hammond's unpublished notes for the Forest in 1937.

In the same year, 1937, Hammond also observed *A. mixta* ovipositing in mud at the edge of a ditch at Benfleet, and H.C. Huggins (1939) reported the species from south-east Essex at about that time. There are specimens collected by D.A. Ashwell in Hatfield Forest in 1940 and by E.S. Brown from Manningtree station in 1947. According to Pinniger's (1942) notes for the Cuckoo Pits survey, the species had occurred with 'some regularity and frequency' in previous years, and that oviposition had been observed. Longfield (1949) still considered *A. mixta* to be scarce as a breeding species, giving Epping Forest as a breeding locality. She also gave Hainault Forest, Ongar Park, Coopersale Common and Chingford as localities, but there is no indication whether these were reports of breeding populations.

Despite the gaps in this historical account, a pattern does emerge. It seems unlikely that *A. mixta* was a breeding species in Essex before the closing years of the 19th century, at which time it appears to have become established on the coast. By the mid 1930s (possibly before) it was established on the Thames estuary as a breeding species, whereas in Epping Forest it probably did not breed until 1936 or 1937, after which time it appears to have bred there in increasing numbers. The spread of records from 1940 onwards suggest that it may also have colonised other parts of Essex at about that time.

Anax imperator Leach, 1815 – Emperor Dragonfly

This common and widespread species is the largest species to be encountered in Essex. It is perhaps the only dragonfly whose English name is in widespread popular usage.

Description

The male emperor has blue-green eyes and a green thorax that contrasts with its sky-blue abdomen. The thorax lacks antehumeral stripes. The abdomen of the female is dull green, and in both sexes there is a black line running along the middle of the dorsal surface of the abdomen. The colours are said to be temperature dependent, with the female abdomen becoming blue in warm conditions, and the male's becoming dirty-green when the temperature cools (e.g. at night) (Brooks (ed.) 2004).

Male emperor dragonfly

Female emperor dragonfly

Similar species

At a distance the emperor might be confused with other large hawkers, especially *A. cyanea*, but the abdominal pattern, lacking the alternating dark bands and paired pale spots of the latter species, is quite distinctive.

The lesser emperor (*Anax parthenope*) is a rare migrant that has been observed twice in Essex. It is smaller than *imperator* and has a brown thorax and, in males, the blue on the abdomen is restricted to segments 2–3. In flight, the abdomen is held in a straight position, in contrast to the down-curved abdomen of *A. imperator*. See also under ***Anax parthenope***, below.

Flight period

The main period of activity is from late May until the end of August. During the current survey, the earliest emergence reported was on 28th May, with a late record of 22nd September. Our impression is that the flight period of this species is more prolonged, both earlier and later, compared with the period of the earlier survey.

Habitat and behaviour

Unlike some other species, *Anax imperator* is rarely seen away from water. They are commonly found hawking above lakes, flooded gravel pits, canals, slow-moving rivers and brackish borrow dykes, the males imperiously patrolling their territories. This patrolling is interspersed with fast, darting flights in pursuit of prey or other male dragonflies that have encroached the boundaries of the defined territory. This territorial behaviour limits the number of individuals that might be found at a single site and D.S. Walker (pers. comm.) observed a successive reduction over several years in the number of males present at a site as reedmace progressively reduced the circumference of the pond. The observers who recorded the second lesser emperor in Essex (at Bedfords Park) considered that it was aggression from the resident emperor dragonflies that hastened the departure from the site of the rare visitor (Jupp & Middleton, pers. comm.).

Patrolling male emperors frequently chase off dragonflies of other species – including other hawkers that encroach on their territories. At a small pond aerial combat between a male *imperator* and a male *A. grandis* appeared to

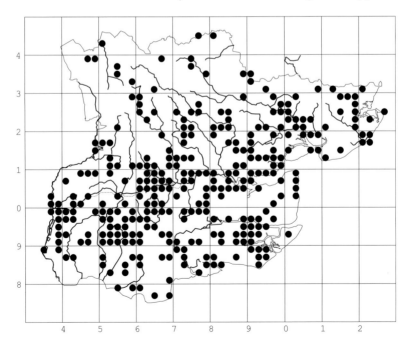

intimidate the latter, but it continued to return to the pond after briefly being chased off (TB, field notes, 09/08/06). Brownett (1998) offers an interesting suggestion that the spacing out of territorial males of this species is one feature (along with nocturnal emergence and timing of the flight period) that makes it less likely to be victim of predation by hobbies than are other species, such as *A. mixta* and *S. striolatum*. Patrolling males are most active around the middle of the day, but also have a further period of activity during the evening.

Airborne prey is relentlessly pursued, sometimes to canopy height, with smaller items being consumed on the wing. Larger items (up to the size of small tortoiseshell and meadow brown butterflies) are usually taken to a perch for consumption. One of us observed a male emperor catch a male migrant hawker (*A. mixta*) in flight, and fly off with it to settle in rank grasses a few metres from the pond (see photograph on p.8). Here it first bit off the head and then ate the thorax and anterior portion of the abdomen. Meanwhile the remainder of the decapitated *mixta* continued to squirm.

Copulating pairs may be found on vegetation up to a few metres away from the water. The female oviposits (without the male in attendance), into the stems of aquatic surface vegetation at sites that are often some distance from the water's edge. The eggs hatch in 3–6 weeks and, due to the long flight period, the larvae may be present at several different stages of development. The larvae live among aquatic vegetation, with different age-groups occupying different niches.

Full development takes one or two years. Larvae emerging after the more usual two years achieve their final instar in the autumn prior to emergence and overwinter in diapause, followed by a synchronised emergence during the following spring. Less developed larvae overwinter in the penultimate instar, with a later, less synchronised emergence the following year (Corbet 1957 & Corbet in Brooks (ed.) 2004).

Distribution and conservation

The emperor is found throughout central and southern Europe, but absent from most of Scandinavia. It extends to Africa, the Near East and parts of Asia. In Britain it is widespread in the south although not recorded from Dartmoor, the Brecon Beacons and chalk downland of Hampshire and Wiltshire. However, there is recent evidence that it is expanding northwards, with regular sightings in Cumbria and the first record for Scotland in July 2003.

The previous survey found the emperor to be 'widespread and common, if somewhat local, across south, central and north-east Essex'. It appeared to be absent from the north-west of the county. At that time it had been recorded from thirty-one 10km squares. Today, *Anax imperator* occurs throughout Essex, including the north-west and has been recorded in fifty-four (of fifty-seven) 10km squares since 2000.

Early records

The emperor appeared (under the name *Anax formosa*) on E. Doubleday's (1835) list for his locality to the east of Epping and J.F. Stephens (1835–7) also gave Epping as a locality. H. Doubleday (1871) described it as 'very common formerly' on Coopersale Common, and recorded it at two large ponds by the 'new road' through the Forest. W.H. Harwood (1903) considered it to be 'now a rarity in the county', quoting Doubleday as evidence that it had once been more frequent. He said 'a single specimen was captured in High Woods, Colchester, several years ago', but no others had since been seen. However, the Campions included it regularly on their lists for the Forest, and it was also listed by E.E. Syms (1929). C.O. Hammond recorded it regularly in the Forest between 1925 and 1945, whilst E.B. Pinniger (1933) noted its 'abundance' there. Longfield (1949) said it was found at 'nearly all' the ponds in Epping Forest and was one of the first species to colonise wartime bomb craters in and around London. Pinniger, Syms and Ward (1950) recorded it from bomb craters on Chingford Plain with further sightings at Monk Wood, Coopersale Common, Benfleet (where Hammond also recorded it in 1937), Hatfield Forest and Hainault Forest, Longfield (1949) also giving Brentwood as a locality for it. There are also two specimens taken by B.T. Ward at Grays in 1950 in the Queen Elizabeth's Hunting Lodge Museum.

Clearly, this species occurred in Epping Forest in the first half of the 19th century with an apparent decline in the late Victorian era, before becoming more widespread in the first half of the last century. It is therefore surprising that the first record for this conspicuous dragonfly in Suffolk occurred as late as 1943 (Mendel 1992), suggesting that the emperor must have previously been scarce in parts of its range.

Anax parthenope (Selys, 1839) – Lesser Emperor

This species is now a regular but uncommon migrant to Britain, with some evidence of breeding.

Description

This is a medium-sized hawker dragonfly, usually slightly smaller than the emperor (*A. imperator*). The thorax is greenish brown, and most of the abdomen is also brownish in males, but is more variable in the female and may be greenish or blue. There is a longitudinal black line down the middle of the dorsal surface of the abdomen (more prominent at the base of the abdomen in females than it is in males), and in both sexes the base of the abdomen is blue, usually with a narrow yellow band.

Some observers report that the abdomen is held straighter in flight than in the emperor, and there is often a yellowish suffusion to the wings.

Similar species

The absence of black and green or blue patterning on the abdominal segments, together with lack of pale thoracic markings, separates this species from the common *Aeshna* hawkers. The main risk is confusion with the far more common emperor (*A. imperator*). The best distinguishing features are:

1. Thorax of *parthenope* is usually brown, but is green/blue in *imperator*.

2. First segment of the abdomen is mainly brown in *parthenope*, but is mainly green in *imperator*.

3. Rest of abdomen is usually brown (but mature females sometimes bluish), but in *imperator* is green becoming blue in mature males, and some females.

Unfortunately there is some variation in colour patterns, and confusion can arise with newly emerged individuals or very old ones. Ideally, all three of the above characters should be checked.

Flight period

In southern Europe the flight season is from March to November, but most UK records are between early June and late August, although there have been occasional sightings up to late September.

Habitat and behaviour

In mainland Europe it is said to favour larger, still-water sites. Unlike *A. imperator*, mated pairs remain together, and oviposition takes place with the pair in tandem.

Status and distribution

After becoming more frequent in north-west Europe during the 1980s, the species was recorded in the UK for the first time in 1996. There have been reports of breeding in the UK, particularly at several sites in Cornwall, since 1999 (Jones 2000). Strong immigrations were noted in 1998, 2003, 2005 and 2006.

Occurrence in Essex:

During the period of our current survey we have only two confirmed records from Essex. These are:

A male at a private former sand-pit near Bradwell-on-Sea (TL9905), seen on 30th July 2000 by M. Telfer, P. Bachelor and others, and reported to the British Dragonfly Society recorder by M. Telfer.

One observed hawking over a fishing lake at Bedfords Park (TQ515920) on 8th August 2003 by C. Jupp and A. Middleton. See a description of the event in **Appendix 4**.

Cordulia aenea (Linnaeus, 1758) – Downy Emerald

This species is the only member of the 'emerald' group of dragonflies to have been recorded in Essex. It always has been, and still is, confined to the west of the county, where it is extremely localised.

Description

The downy emerald is a medium-sized dragonfly with a dark, metallic green abdomen and thorax, with buff-coloured hairs on the thorax (hence 'downy'). The eyes are of a brighter green tint, and there are pale yellowish to whitish markings on the sides and ventral surfaces of abdominal segments 1 and 2. The abdomen of the male is constricted at segment 2, and broadens out towards the rear end, giving a distinctive club-shaped appearance. The abdomen

Downy emerald in flight

of the female is more parallel sided, but she is otherwise very similar to the male.

Similar species

The only species with which this could be confused are the other two British emeralds: the brilliant emerald (*Somatochlora metallica*) and the northern emerald (*Somatochlora arctica*). However, neither of these is likely to be found in Essex, and the northern emerald, in particular, has no known English localities. The brilliant emerald has a distinctive yellow band on the front of the face, and the anal appendages are rather longer than in the downy emerald.

Downy emerald at rest

Flight period

During the 1980s this was reported to be from the end of May through to the end of July (Benton 1988). During the current survey earlier spring emergences were noted, the earliest on 14th May. The latest date reported was 25th July.

Habitat and behaviour

The downy emerald breeds almost exclusively in shallow, sheltered, ponds, lakes or canals in woodland. These are usually neutral or slightly acidic, and

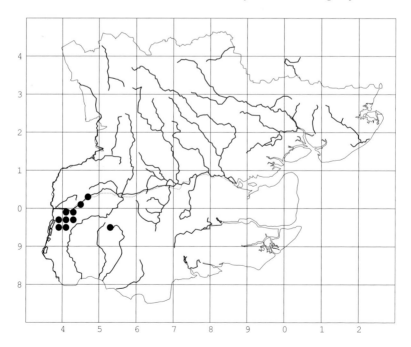

poor in nutrients. Although closer in size and general appearance to 'darter' dragonflies, downy emeralds behave more like hawkers. The males patrol territories close to the margins of the breeding sites, flying quite low, and hovering frequently. The patrolling continues even in quite overcast weather conditions. They rarely settle on the pond margins, but after a period of up to a half-hour (average recorded 12.75 minutes, see Brooks, McGeeney & Cham 1997), disappear into surrounding woodland, only to reappear later to continue their territorial behaviour.

Through the day several different males may 'time-share' a given section of a pond margin, but males arriving at an already-occupied territory are chased off by the incumbent male, as are intruding 'chaser' (*Libellula* species) dragonflies. Passing females are intercepted and grasped, the pair then flying in tandem up to the tree canopy, where mating takes place. The females return to lay their eggs close to the water margin, usually avoiding very shaded areas. They repeatedly flick the tips of their abdomens below the water-surface, and lay their eggs in small batches (reputedly 10 at a time (Brooks 2004)). Between bouts of patrolling (males) or egg-laying (females), the dragonflies may be seen hawking for insect prey along woodland rides or in open glades.

The larvae live among dead leaves and other detritus at the bottom of their ponds, hanging upside-down from larger pieces of leaf. They feed at night on small invertebrates such as water-lice, and the aquatic larvae of small insects such as midges and alderflies. The larval development is slow, taking two or more years. When ready to emerge as adults the larvae climb up the stems of marginal vegetation. An interesting experiment has demonstrated that prior to emergence from the final larval 'skin', the larva makes circular movements with its hind legs to ensure that there is sufficient space around it for its wings to expand. If a physical obstacle is encountered, the full-grown larva moves off and finds another place to complete its metamorphosis (Wildermuth 2000). As soon as the wings of the newly emerged adults have hardened they fly directly up into the tree canopy above.

Distribution and conservation

The downy emerald is widespread in central and northern Europe, but is absent from the southern countries and northern Scandinavia. In Britain it is most frequent in the south-east, but has an oddly discontinuous distribution further north including some parts of Scotland. Its strongholds are in the wet heaths of Surrey, Hampshire and Dorset. During the 1980s, E.P. Ryan and others observed it regularly at Earls Path Pond, Lower Forest Lake, Wake Valley Pond and the Lost Pond, all in Epping Forest (Benton 1988). Repeated visits to Hatfield Forest were unsuccessful in confirming the continued existence of the species there. During the period of the present survey, the species has been observed to occur more widely in the Forest than reported in the earlier study, but, again, it has not been found at Hatfield Forest. There is one sighting away from the Forest, almost certainly not representative of another breeding population. However, this does indicate considerable powers of dispersal and suggests that future colonisation of suitable sites should not be ruled out.

The downy emerald population in Epping Forest seems secure so long as management remains appropriate. Breeding during the period of the current survey has been confirmed at Wake Valley Pond, Little Wake Valley Ponds east and west, Staples Road Pond, Baldwin's Hill Pond, Lost Pond, Earl's Path, Fairmead and Goldings Hill ponds (A. Samuels pers. corr.). It seems probable that it also breeds in Lower Forest Lake. What is known about its ecology suggests that drastic clearing of adjacent trees and dredging of the pond floors should be avoided.

Early records

It is likely that the downy emerald has bred in the Epping Forest ponds continuously since our earliest records. According to Edward Doubleday (1835) it occurred 'in profusion', and Epping was also given as a locality for it by Stephens (1835–7). According to Henry Doubleday it was 'very common' on Coopersale Common and at other places near Epping. The Campion brothers recorded it in the Forest each year from 1906 to 1909. It appears on two lists for the Forest during the 1920s (Roberts, cited in Lucas 1927, and Syms 1929), and was reported to be common by Pinniger from 1933 onwards (Pinniger 1933, 1934a, 1935, 1936, 1937). Hammond noted it on numerous occasions between 1923 and 1945, and Longfield (1949) said it was to be seen at all the larger ponds in the Forest. In the Queen Elizabeth's Hunting Lodge Museum there is a specimen collected by B.T. Ward in 1949 from the Forest, and Pinniger, Syms & Ward (1950) included it in their 1949 list. Outside Epping Forest, we know of only two early localities: one at Woodford by Harcourt Bath, cited by Harwood (1903), and the other the lake at Hatfield Forest, where it was recorded by Ashwell in 1939 (BRC), and a specimen collected in 1940 (now in the Natural History Museum, London).

Libellula depressa Linnaeus, 1758 – Broad-bodied Chaser

This medium-sized dragonfly is widely distributed in Essex and is most frequently observed at shallow ponds and lakes.

Description

Both male and female have a broad, flattened abdomen, which is powder blue with yellow lateral spots in mature males and brown with yellow margins in the females. In both sexes the thorax is brown with a pair of pale antehumeral stripes. There is a dark brown triangular patch at the base of each hindwing, and also some dark suffusion near the base of each forewing.

Similar species

The females can be distinguished from other species by shape. The males of two other Essex species could be confused with the males of *depressa*. These are the scarce chaser (*L. fulva*) and the black-tailed skimmer (*O. cancellatum*). Males of both these species have pale blue abdomens when mature, but both are narrower in shape. The male scarce chaser lacks the yellow patches on the sides of the abdomen, has smaller dark patches on the wing bases, and has a black tip to the abdomen. The scarce chaser male has blue eyes (brown

Male broad-bodied chaser

Female broad-bodied chaser

in the broad-bodied). Mature male black-tailed skimmers also have black tips to the abdomen and lack the dark wing-bases. The very rare visitor, *O. coerulescens,* is also blue in the male, but has a much narrower body in that species, and lacks dark wing-bases.

Flight period

The flight season in Essex was reported to be from late May to mid-August in the 1980s. During the current survey we have earlier records (earliest 11th May), with an exceptionally late record of a worn female at Harold Court on 11th September 2002 (C. Jupp, pers. comm.).

Habitat and behaviour

This species is generally recorded from small bodies of water such as ponds (often in gardens or woodland) and dykes with abundant marginal plants. As these ponds become more choked with vegetation, they are less attractive to this species, which, together with *O. cancellatum,* is able to colonise newly created sites and would appear to be a beneficiary

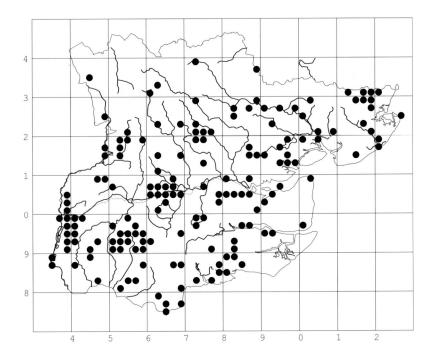

of the increase in ponds and farm reservoirs that has occurred during the last twenty years.

Females are often found perched on vegetation away from breeding sites whilst the males are strongly territorial at the ponds where they breed. They survey their territory from a prominent position, flying out to grasp passing females or to chase off intruders. Mating is brief , usually accomplished in flight, and the female begins ovipositing immediately afterwards. The male 'stands guard' at this time and may occasionally attempt to mate again with the female. The larvae live among fine detritus at the bottom of their pond, and are able to survive in damp mud during periods of drought. Larval development is prolonged, taking from one to three years.

Winsland (1995) reports predation on *L. depressa* during emergence from the larval stage by black ants (*Lasius niger*), and there is an account by Cross (1987) of predation by a pied wagtail. Apparently the bird showed no interest in patrolling males, but attacked less manoeuvrable pairs as soon as a male was able to grasp a female in the air. The bird took only the female dragonfly, and took it to the bank where it removed the wings and consumed the body.

Distribution and conservation

The broad-bodied chaser is widespread in southern Britain, although absent from most of the north. However, recent records from Darlington in 1999, Spurn National Nature Reserve and a site near Scarborough in June 2000 (Parr 2001) suggest that the species may be extending its range. It may be found throughout Europe north to southern Scandinavia.

The current survey found *L. depressa* more widely dispersed than previously. It was recorded from forty-seven 10km squares compared with twenty-seven during 1980–87 and this increase may be attributed to the increase in suitable water bodies (except in the north-west of the county, where it remains scarce) that has occurred recently.

Early records

Henry Doubleday's Epping List (1871) described the broad-bodied chaser as 'very common' and F.A. Walker recorded it at Wanstead Park in 1897. W.J. Lucas (1900b and 1902b) included records from Maldon, Colchester and Navestock, while Harwood (1903) described it as 'generally distributed and usually common'. The Campion brothers (1909) reported it as common on the shallow ponds in the Forest during the early years of the 20th century and, interestingly, included a record of an individual on the wing as late as 7th September 1908. E.E. Syms (1929) included it in his list for Epping Forest and E.B. Pinniger (1933) reported it to be 'fairly common each year' in the Forest. Hammond also regularly observed *L. depressa* in the Forest during the 1920s while Longfield (1949) described it as well-distributed in the LNHS area and as more abundant than its relative, *L. quadrimaculata*. Pinniger, Syms and Ward (1950) added sites at Wanstead Park, Chingford Plain, Stanford Rivers, Margaretting and Childerditch Common where the species was observed in 1949, and other scattered records from Grays, Althorne, Fingringhoe Wick and Connaught Water are documented between 1950 and 1980.

Libellula fulva Müller, 1764 – Scarce Chaser

One of the highlights of the current survey has been the confirmation that the scarce chaser, not recorded in the previous survey, is now present on three of the river systems in the county.

Description

Females and immature males have bright, orange-brown abdomens with a distinctive longitudinal black line down the middle of the dorsal surface. This broadens out to form roughly triangular markings from segment 4 to the tip of the abdomen. In the female this pattern is retained in full maturity. She also has a dark, smoky-coloured tip to the wings. The mature male has pale-blue pruinescence on the dorsal surface of the abdomen, with a black tip. Both sexes have dark brown triangular patches at the base of the hind-wings, and some slight darkening of the basal area of the forewings. There are no antehumeral stripes.

Male scarce chaser

Similar species

This species is one of four British dragonflies where the mature male has a blue colouration on the abdomen. The mature males of *fulva* are very similar to those of *O. cancellatum* but they can be distinguished by the dark bases to the hindwings – clear wings in *cancellatum*.

Some care is needed in using this character as the dark patches in *fulva* are rather small and not always easily seen when the insect is viewed from the side. Males of *L. depressa* have much broader abdomens, and yellow patches

at the sides of the abdominal segments. They also have pale antehumeral stripes. The very rare (in Essex) keeled skimmer (*O. coerulescens*) also has a blue abdomen in the male, but lacks the black tip to the abdomen and the dark patches on the wing bases. The patterning of the female abdomen in the scarce chaser is distinctive. In addition, her abdomen is narrower than that

Female scarce chaser

of the broad-bodied chaser, and she lacks the dark spots at the wing-nodes of the four-spotted chaser. The dark patches at the wing bases distinguish her from both of the 'skimmer' (*Orthetrum*) species.

Flight period

The earliest recorded date for the scarce chaser during the current survey was on 29th May 2005, when seven or eight teneral specimens were observed on bankside vegetation along the Chelmer/Blackwater Navigation near Langford Golf Club (R. Neave, pers. comm.). The latest date recorded was the 12th July (in 2003).

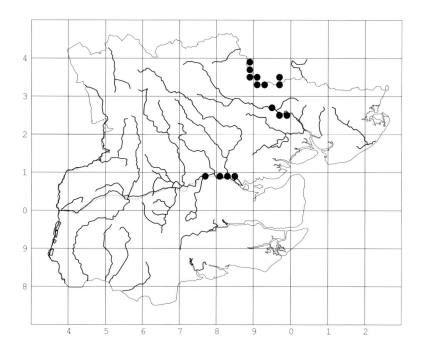

Habitat and behaviour

The scarce chaser inhabits river floodplains, water meadows and their associated ditch systems, but has occasionally been recorded from mature gravel
pits – especially when these are close to rivers where they breed. Typically,
slow-moving reaches of rivers and canals are used as breeding sites. These
tend to have dense emergent and bank-side vegetation, and to have bushes
or trees close to the river-bank on one side of the river. Often these stretches
are characterised by stands of bulrush and arrowhead emergent from mid-
stream, and males frequently use the stems of the former plant, as well as
reedmace, as perches. They, along with other firm emergent plants such as
common reed, are also used as supports for emergence.

Immature specimens may often be seen close to the breeding sites, but usually
take their first flight up to the tree canopy, when they are vulnerable to predation by birds (see **Chapter 1**). The mature males take up territories close to the
river bank around noon, spending much of their time on high perches, but
flying off to chase away intruding males or intercept females. Mating pairs
settle in dense vegetation near the breeding site and after mating the female

lays her eggs by 'flicking' the water surface with the tip of her abdomen. The male is usually present, keeping guard nearby. Late in the season the male abdomen becomes discoloured with black markings caused by the grasping of the females during mating, and the wings become tattered. Over-night they roost very low down in dense vegetation, climbing up to warm themselves in early sunshine before taking their first flight of the day (Cham 1999).

The larvae live among decaying plant material and roots, and take two years to complete their development.

Distribution and conservation

L. fulva is locally common, but with a rather patchy distribution, through central and eastern Europe. In England it is near the northern limit of its range, and it is localised in the south-east. It was once thought to be restricted nationally to six distinct areas along the rivers Yare in Norfolk and Waveney in Suffolk; the Avon in Wiltshire/Somerset; the Arun in Sussex; the Nene and the Ouse in Cambridgeshire; the North Stream and associated ditches in North Kent and the Frome, Stour, Moors and Avon in Dorset/Hampshire. There is now evidence that this species is currently expanding its range, with recent reports of sightings coming from the River Wey in Surrey (Follett 1996), the Soke of Peterborough (Tyrell & Brayshaw 2004), several sites between Pershore and Tewkesbury on the River Avon (Worcestershire, BRC 2004) and the Stour, Colne and Chelmer/Blackwater in Essex.

The recent discovery (or rediscovery) of this species along the Stour has been reviewed by Cham (2000). Ian Johnson first reported two males in July 1997 as he led a field trip near Nayland and further sightings were by Cham himself who recorded two more males near Bures in 1998. In the following year, a sizeable population with evidence of breeding was recorded between Bures and Nayland. More recently, the species has extended further along the Stour and been recorded at Wormingford, Lamarsh and Henny Street.

Away from the Stour, T. Gunton (pers. comm.) reported a sighting in south-west Essex in 1997, and the first sightings of the species on the River Colne (near Earls Colne) were reported in 1999 (Benton & Dobson 2002).

Although the Stour may have been colonised naturally (or the species may have been overlooked) Cham suggests an alternative possibility. Mendel (1992), in seeking to explain the increase in *E. najas* along the Stour, noted that large volumes of unchlorinated water are now regularly pumped underground from Denver on the Ely Ouse into the upper reaches of the Stour. He speculates that eggs and larvae of *L. fulva* – and *P. pennipes* which is also more common on the Stour than previously – could have been introduced in this way. It is a plausible explanation, although increases in the distribution of the latter two species have also been seen elsewhere in Essex.

Ted Benton first searched the lower reaches of the River Colne in June 2000 and observed two or three specimens of *L. fulva* close to a sheltered bend in the river. Unfortunately the site was inaccessible due to foot and mouth restrictions in 2001. Despite searching, none were seen in 2002, and it was not until 21st June 2003 that the species was again recorded from the same area. During the same year, the species was first found to have colonised the Chelmer/Blackwater Navigation with several being found along the stretch of water close to Langford Golf Club. Further searching produced other sites, the most westerly being near Papermill Lock, Little Baddow.

In Essex, records have now been obtained from the Colne near Earls Colne, at West Bergholt and alongside Cymbeline meadows to the west of Colchester, the River Stour between Little Henny and Nayland, and in four tetrads from the Chelmer/Blackwater Navigation. It will be of interest to see if the scarce chaser is able to extend its range in the county in the future. There are further stretches of the Colne, Stour and Chelmer/Blackwater that look as suitable as those that are currently occupied and the Lea Valley also shares many of these features.

Early records

This is a species that has justified its English name in Essex. Doubleday (1871) referred to it as 'rare': it had occasionally been seen flying over a pond in Ongar Park Woods. Lucas (1900) confirmed Doubleday's observations and also added a further sighting by Harwood at Colchester, which was presumably the same specimen referred to in Harwood's (1903) list. There are no further reports for Essex until Longfield (1949) referred to a record from Colchester in 1905 (presumably a misdated reference to Harwood's report). Longfield also referred to a colony on the Suffolk Stour although Mendel (1992) in his review of Suffolk's dragonflies concluded that this record was best regarded as 'unconfirmed'. Intriguingly, Tony Gunton (pers. comm.) reported seeing this species in the south-west of the county in 1997 – the first confirmed record for almost a century.

Libellula quadrimaculata Linnaeus, 1758 – Four-spotted Chaser

Although considered scarce in the county only twenty years ago, this me-dium-sized dragonfly is now widespread in most of Essex.

Description

The males and females of this species are very similar, and the males, unlike their relatives in the genus *Libellula*, do not acquire the blue colouration on the abdomen. Both males and females have orange-brown abdomens, with a darker tip. They become darker brown with age, when the narrow yellow patches at the side of each segment become more clearly visible. There is a dark brown triangular patch at the base of each hindwing, and a dark spot at the node (mid-point of the leading edge) of each wing (hence 'four-spot-ted'). There are no antehumeral stripes. A very striking form with additional dark patches at the wing-tips is termed f. *praenubila* (Newman), and was considered by both Edward Doubleday (1835) and Edward Newman (1832–3) to be a separate species, until Vander Linden's view that it was merely a variety of *quadrimaculata* prevailed.

Similar species

Both sexes of the four-spotted chaser may be distinguished from similar species by the presence of the distinctive dark spot at the node of each wing.

Male four-spotted chaser

This feature separates it from females or immature males of *O. coerulescens*, *O. cancellatum*, *L. fulva* and *L. depressa*. See under those species for further detail.

Flight period

This species is one of the first dragonflies to emerge during the spring, with the earliest date for the current survey being 7th May. This is a full month earlier than the earliest date recorded in the previous survey, but the species was more scarce at that time and the full flight period may not have been recorded. Typically a spring species, individuals can still be seen well into August (15th August the latest date reported during the current survey).

Habitat and behaviour

In Essex, the four-spotted chaser is not only found on ponds, canals, slow-flowing rivers, gravel pits and farm reservoirs, but also on brackish dykes near the coast. At all sites, there is a requirement for well-established emer-

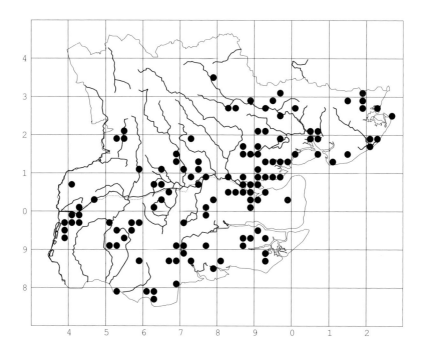

gent vegetation for the males to use as a favoured perch, from which they make regular sorties to see off rival males or intercept females. There is often intense competition for perches, and the authors observed repeated conflict between a male of this species and a male *L. depressa* over occupation of a prominent perch overlooking a small pond (TB, field notes, 05/06/06). Where there is high density of males at a pond (as observed, for example, at some of the woodland pools at Marks Hall), territorial behaviour seems to break down, and the males spend much of their time on the wing and in aerial chasing behaviour. Females are often found away from the breeding pond, perched on bushes and shrubs, but return to the water to mate and lay their eggs. Mating occurs in flight and is over in 5–20 seconds (Brooks (ed.) 2004). After copulation, the female oviposits by hovering over the water (usually at a site with submerged vegetation), repeatedly dipping the tip of her abdomen into the water to release the eggs. The male is usually in attendance nearby. Like other 'chaser' larvae, those of *quadrimaculata* live among detritus near the pond floor, and development takes two years.

Mass adult emergence has been recorded and, where the species is very common, large migrations may be seen. To give a celebrated example, Hagen records that in Konigsberg in 1852 he saw 'a cloud 500m long, 20m wide and 3–4m high of serried ranks flying slowly in the same direction' (Aguilar *et al.*, 1986). Such a swarm must have involved millions of individuals.

Benton (1988) reports predation on this species by spiders, and males are frequently taken by emperor dragonflies (Brooks 2004). Bailey (2000) also gives otter as a predator on this species, as evidenced by clusters of severed wings. Adult *quadrimaculata* will catch quite large insect prey – including bumblebees (Radford 2000).

Distribution and conservation

This species is common throughout most of Europe, but is absent from much of the south and from northern Scandinavia. Its dark colouration, fine hairs on the thorax and transparent cuticle make it well-adapted for cooler regions, and it is widespread throughout the British Isles, although less common in north-east England.

The historical record for this species shows considerable fluctuations from year to year, presumably as a result of migrations, which Dumont and Hinnekint (1973) suggest occur cyclically with a periodicity of about ten years.

Between 1980 and 1987, the species was recorded from just fourteen sites in ten 10km squares, with good evidence of breeding only at rather few of those. By contrast, during the present survey, it was found at 110 sites in thirty-four 10km squares. This increase may in part be due to immigration but it has also coincided with an increase in recorder activity and to a greater number of suitable water bodies being created, such as farm reservoirs. The apparent absence of this species from north-west Essex is considered to be due to a lack of suitable habitat rather than a lack of recorder effort. It is possible that the reported preference of *quadrimaculata* for acidic water bodies is relevant here, given the geology of that part of Essex (former chalk downland).

Early records

The four-spotted chaser featured on the early lists for the Epping area (E. Doubleday, 1835) Stephens, (1835–7) and H. Doubleday (1871), and Newman referred (1832–3) to *Libellula praenubila* as taken by 'my friends H. & E. Doubleday' at Epping. At that time, H. Doubleday declared *L. quadrimaculata* to be the commonest species of the family.

It also appeared on F.A. Walker's (1897) list for Wanstead Park and W.H. Harwood (1903) added Wivenhoe (for one day only in 1900), Clacton (where the variety f. *praenubila* was recorded) and Colchester as new sites – presumably the results of immigration. He also mentioned a large migratory 'swarm' of *L. quadrimaculata* seen off the Essex coast in June 1888. *Libellula quadrimaculata* appeared on Epping Forest lists subsequent to those of the Doubledays, but the Campion brothers (e.g. 1906 and 1908) commented on its scarcity in the Forest. E.E. Syms (1929) listed it, as did E.B. Pinniger (1933), who recorded it as common in favourable years (possibly as a result of immigration).

C.O. Hammond recorded it in the Forest in the 1920s, noting in 1924 that it was 'common at all ponds' and continued to record it up to 1945. Longfield (1949) expressed surprise at its local status and scarcity in the London area, but went on to point out, however, that it had apparently increased as a breeding species in the Forest. She gave 'muddy little ponds', gravel and brick pits, as well as fresh and brackish marshes, as breeding sites for the species.

Orthetrum cancellatum (Linnaeus, 1758) – Black-tailed Skimmer

Although first recorded in Essex as recently as 1934, this dragonfly is now a familiar sight around ponds and lakes throughout the county.

Description

Mature males have a brown thorax, grey-blue eyes and a pale blue pruinescence on the abdomen. There is a black tip to the abdomen, and narrow yellow patches at the sides of each segment. Immature males and females have yellow abdomens with two black bands running along the dorsal surface. As the male approaches maturity, the blue colouration gradually obscures the immature pattern. The wings are clear, lacking dark patches near their base, and there are no antehumeral stripes.

Similar species

The yellow-and-black abdominal pattern distinguishes females and immature males from all three of the 'chaser' (*Libellula*) group. Immature males of the keeled skimmer (*O. coerulescens*) can look quite similar, but they are

Male black-tailed skimmer

Female black-tailed skimmer

smaller and more delicately built – and very unlikely to be seen in Essex! The mature males could be confused with the males of three other species. The most likely confusion is with the scarce chaser (*L. fulva*), as the two species can be seen together in a few places.

The most reliable distinguishing feature is the lack of the brown triangular patch at the base of the hindwings in *cancellatum* (present, but rather small and inconspicuous in *fulva*). *L. fulva* also lacks the yellow sides to the abdominal segments that are present in *cancellatum*. The broad-bodied chaser (*L. depressa*) has a much wider, flattened abdomen, and a pair of antehumeral stripes on the thorax (absent in *fulva*). The dark patches at the base of the hindwings are also well-developed and conspicuous in *depressa*. Finally, mature male *O. coerulescens* lack the black tip to the abdomen, and have antehumeral stripes

Flight period

The flight period is from late May to the end of August, with peak numbers being recorded in July. There seems to have been little change since the previous survey. During the current survey, the earliest sighting was on 24th May and the latest observation – an exceptional record – on 21st September.

Habitat and behaviour

The black-tailed skimmer is commonly found on flooded sand and gravel workings, farm reservoirs and other ponds and lakes. Not requiring much marginal vegetation, it is one of the first colonisers of new ponds and (along with other species) has been a beneficiary of the increase in wetland creation that has recently occurred in Essex. The males commonly bask on bare ground and, as banks become more densely vegetated, so ponds and lakes become less suitable for this species.

However, the widespread use of water bodies for angling has provided a combination of dense bankside vegetation, alternating with bare ground or wooden decking as fishing pegs, thereby ensuring that ideal basking conditions remain as the lake matures. *O. cancellatum* is also found in coastal areas, arriving soon after the excavation of new borrow dykes and ditches,

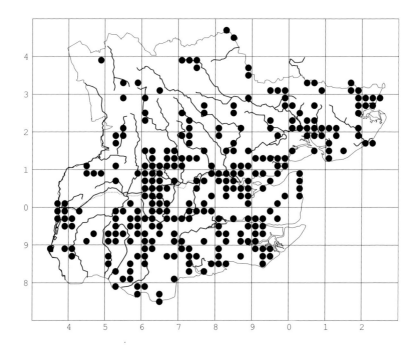

and remaining to breed in them and other brackish pools, long after the vegetation has become established. It may also be found on slow moving-water sites with territorial males being recorded along the River Colne, the Chelmer/Blackwater Navigation and the Cornmill Stream.

Mature males acquire a territory along a length of water's-edge and are frequently on the wing, patrolling the prospective breeding site and reacting aggressively towards other males. In this respect, they resemble hawkers in behaviour and, in their turn, are often victims of aggression from larger hawkers (notably *Anax imperator*) on adjacent territories. Perhaps as an adaptation to this aggression, they characteristically fly close to the surface of the water and marginal vegetation, in contrast to hawkers, which usually tend to fly higher and further out.

Males of *O. cancellatum* also exhibit darter-like behaviour, which consists of perching, wings down-swept, on a reed or stem at the water's edge from where regular sorties are made before returning to the same perch. Before mating, females are infrequently seen close to water, preferring to feed – along with immature males - in open areas nearby. On returning to the pond, they are seized by territorial males, with copulation lasting 20–30 seconds and typically taking place in flight, often while the pair hover close to marginal vegetation. The previous survey (Benton 1988) recorded mating pairs being found on bare mud or gravel (usually late in the day) with copulation being longer in duration. These longer copulations (lasting up to fifteen minutes) include the removal from the female of any sperm from rival males, prior to the male fertilizing the female with his own sperm (Miller in Brooks (ed.) 2004).

Egg-laying usually occurs immediately after mating, the female repeatedly dipping the tip of her abdomen into the surface of the water. At this time, the male hovers in close attendance or sometimes perches nearby. The eggs hatch after 5–6 weeks, and the larvae (like those of the 'chaser' species) live in leaf litter and other debris. They take up to three years to develop.

Distribution and conservation

The black-tailed skimmer is found throughout Europe but is absent from most of Scandinavia and northern Britain. Its global distribution includes North Africa and Asia, extending to northern India. In Britain, it was formerly confined to southern England and parts of Wales but recent records from Pembrokeshire, Lancashire and County Durham confirm that it is extending its range. It is unclear how far this extension of range is connected to the creation of many new suitable habitats in the shape of mineral extraction pits, and how far it is an indicator of climatic change.

Between 1980–87, *O. cancellatum* was described as having continued its expansion and was 'well-established and generally common' in several parts of the county. At that time, it was recorded from fifty-four tetrads in twenty-four 10km squares. Since then, the expansion has continued further, with the current survey recording the species in 276 tetrads in fifty 10km squares.

Early records

O. cancellatum was first discovered in Essex at Hainault Forest by E.B. Pinniger on 4th August 1934 (Pinniger 1935), who later confirmed that the species bred there by collecting a nymph from the site (1936). Coincidentally, the first Suffolk record was also around this time with a specimen observed on 3rd June 1935 (Mendel 1992). Hammond's notes record that he observed it on one occasion in Epping Forest in 1937, but it was still regarded as rare and a new species to the Forest in the report on the Cuckoo Pits Survey (Pinniger 1945). Pinniger also suggested that *O. cancellatum* might be breeding in nearby bomb craters – an idea that seems likely, given this species ability to colonise newly created sites. Longfield (1949) gave Hainault Forest as a locality but commented that for 'some unexplained reason' it had only been recorded in Epping Forest on two occasions.

However, a specimen collected by E.S. Brown from the Forest in 1947 in the Hope Department collection, together with Pinniger's and Hammond's records, suggest that it was more frequent in the Forest by the end of the 1940s than Longfield thought. Pinniger, Syms and Ward (1950) reported sightings in 1949 of the species from Hatfield Forest (where it had also been observed by D.A. Ashwell) and from two separate sites in Epping Forest. There are also specimens in Queen Elizabeth's Hunting Lodge Museum collected by B.T. Ward from Hatfield Forest in 1949 and Grays in 1950. During the 1970s, there were further records from the Forest by J. Owen Mountford, Hammond and others but generally, between 1950–1980 there is a dearth of records due more to a lack of observers rather than a decline in the species. As the results of the last two surveys show, this species has expanded its status and distribution in Essex during the last seventy years, and is now widespread throughout the county.

Orthetrum coerulescens (Fabricius, 1798) – Keeled Skimmer

Description

Mature males have a narrow, uniformly pale blue abdomen, blue eyes, and grey or grey-blue thorax. There is a pair of pale antehumeral stripes on the thorax. The females are yellow-ochre, with paler antehumeral stripes, and a yellowish tint to the wings. Their eyes are greenish.

Similar species

Mature males can be distinguished from *O. cancellatum* and the blue *Libellula* species by a combination of the narrowness of the abdomen and lack of a black tip. Immature males (most unlikely to be seen in Essex) could be confused with females or immature males of

Male keeled skimmer (A. Middleton)

O. cancellatum. However, the presence of antehumeral stripes and the smaller, more delicate build of this species should distinguish it. Females of *coerulescens* could be confused with female *Sympetrum striolatum* or *sanguineum*, but they lack the black markings on the rear segments of the abdomen that are present in those species.

Habitat and distribution

Widespread in mainland Europe, but more localised towards the north, *coerulescens* is less specialised in its habitat requirements than in Britain. Here it favours wet heathland, sphagnum bog and moorland, and its distribution in Britain is constrained by the presence of these habitats: the lowland heaths of Hampshire, Dorset and Surrey, the south-east peninsula, west Wales and western Scotland are its main UK strongholds. There is a population at Holt Lowes in Norfolk.

Occurrence in Essex

A single male was taken by the Campion brothers in Epping Forest near Chingford on 22nd July 1900 (Campion & Campion 1906b). It is surprising that the species did not appear on earlier lists for the Forest, as many of the other species of wet heathland that fly with *O. coerulescens* elsewhere were formerly present. Small areas of potentially suitable habitat have been re-created in recent decades, but the Forest is a long way from other breeding populations, so that natural colonisation might seem unlikely, and there are no substantial areas of suitable habitat elsewhere in the county. However, there are occasional reports of individuals of this species seen at consider-able distances from their known breeding sites – e.g. at Bishops Stortford, very close to the Essex border, and at several places in Norfolk in summer 2003 (Parr 2004).

We have received three recent records. The first, just outside the survey pe-riod, was from A. Middleton. This was a sighting of a single individual at Long Running, in Epping Forest on 8th and 9th July 1999. The second re-port, from D. Blakesley, was, again, of a singleton. This was observed on 20th July 2006 at the entrance to a new Woodland Trust site at Theydon Bois (TQ458984). This is, of course, very close to Epping Forest, and suggests that the species is at least able to reach potential breeding sites in the Forest. There is also a report of another seen in the Epping Forest/Lea Valley area on 28th July 1999 by B. Wurzell (Parr 2000).

Sympetrum sanguineum (Müller, 1764) – Ruddy Darter

This medium sized, bright red darter appeared to be in decline by the 1970s, but has become more common and widespread since that time.

Description

Fully mature males have red-brown eyes, a brown thorax and a bright red abdomen. There are black markings at the sides of the abdomen, and a median black line on the dorsal surface of the abdominal segments 8 and 9. The abdomen is club-shaped, due to a slight narrowing of the middle segments. The legs are uniformly black, and there is a slight orange-yellow tint at the base of each wing. The fine black bar across the face is continued downwards at the sides,

Male ruddy darter

resembling a droopy moustache. The females have more parallel-sided yellow-ochre abdomens, with a similar pattern of small black markings to that of the males. The legs are black.

Similar species

The ruddy darter is very similar to the common darter (*S. striolatum*), and the two often occur in the same localities. The most useful characteristics for field identification are:

1. The club-shaped abdomen in male *sanguineum* (roughly parallel-sided in *striolatum*, but this feature is clearest from above. In side view the abdomen of *striolatum* is somewhat club-shaped, too)

2. The black legs of both sexes of *sanguineum* (brownish with a fine yellow stripe in *S. striolatum*).

3. The 'droopy moustache' marking on the face of *sanguineum* (black bar not continued down the sides of the face in *striolatum*)

4. The prominent vulvar scale on the ventral surface of the abdomen towards the rear end (side-view) in the female *striolatum* (inconspicuous in the female *sanguineum*).

Other similar species are much less likely to be seen in Essex. The female of the black darter (*Sympetrum danae*) is similar, but has a distinctive black triangular marking on the dorsal surface of the thorax (see **Appendix 1**). For distinction from the uncommon migratory 'darters' these species, see under *S. fonscolombii* below, and under *S. vulgatum* (**Appendix 1**).

Female ruddy darter

Flight period

This species is most common during July and August, but during the current survey was recorded as early as 15th June, and as late in the Autumn as 11th November. Compared with the previous survey-period, this suggests a considerably extended flight period, with emergence some two weeks earlier, and survival of some individuals much later into the Autumn.

Habitat and behaviour

This species inhabits a wide range of (mainly) still-water sites, typically breeding in shallow, more densely vegetated water bodies than its relative *S. striolatum*. It is abundant in coastal and estuarine borrow dykes and fleets, where it breeds in shallow and often brackish ditches even where these are choked with sea club-rush. Here, it is one of the few species that breeds in the same ditches as the scarce emerald damselfly, *L. dryas* (Thomas 1999).

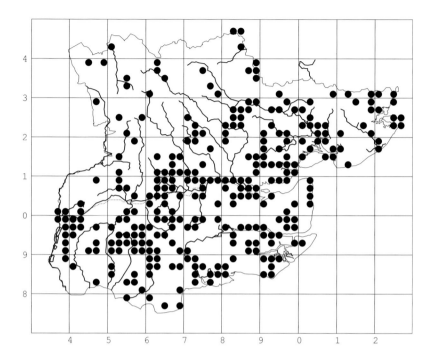

It is also frequently to be seen at mineral extraction pits, especially where these have developed marginal and aquatic vegetation, forest and woodland pools, garden ponds and, less commonly, canals and slow-flowing reaches of rivers.

Freshly emerged specimens disperse some distance from the breeding site, where they adopt perches, usually at the tip of a twig, or on top of a tall herbaceous plant. They fly out to catch prey or chase off interlopers, usually returning to the favoured perch. When mature they approach a suitable breeding site. Males again establish loosely defined territories, often selecting one or more favoured 'perches', but also spending much of their time on the wing.

When at rest on the perch, the wings are held pointing downwards and forwards, but there is another posture adopted when first landing on a perch and at other times. In this, the abdomen is raised and the wings spread out flat. This is interpreted as a threat posture (McGeeney in Brooks (ed.) 2004), but may also be a more generalised indication of readiness to fly.

Passing females are grasped by the male, and the pair land in nearby grass tussocks or other dense vegetation during mating. After several minutes the pair switch to the tandem position, and seek out a suitable site for egg-laying. This is usually close to a pond margin, over superficial vegetation or among emergent plants, but is sometimes over damp mud.

Egg-laying is accomplished by the male swinging the female forwards to 'swipe' the water or other surface with the tip of her abdomen as she releases a batch of eggs. This is repeated in a regular rhythm as the pair work their way along a pond margin. Occasionally females can be seen laying their eggs 'solo', or with a male hovering close by.

The eggs hatch in a few days unless laid late in the year, in which case there is a resting phase ('diapause') and hatching is delayed until the following spring. The larvae live among roots, usually close to the edge of the water-body, and take approximately one year to develop.

Taylor (1994) reports predation on an individual of this species, taken in flight, by the wasp *Vespula germanica*.

Distribution and conservation

The ruddy darter occurs throughout most of Europe except for the Alps and some parts of the extreme south and south-west. In Scandinavia it is restricted to the south, and also has a mainly south-eastern distribution in Britain. However, it has colonised much of Ireland since 1920, and in recent decades has spread further north and west in Britain.

After a period in the 1970s when alarm was expressed about an apparent decline in this species (along with the feared extinction of *L. dryas*) it recovered strongly. By the time of the previous survey it was recorded from forty-two 10km squares in the county, with a greater concentration in the east and south, and with very scattered records away from its eastern strongholds.

As noted at the time, although breeding in many sites was well-established, numbers were augmented each year by migration. Some twenty years later, the species seems to be much more firmly established as a breeding species across the whole county, and our subjective impression is that it may now be as common and widespread as the common darter. In the current survey it has been reported from fifty-three 10km squares across the county.

Early records

H. Doubleday (1841) expressed puzzlement about this species' (as 'S. *ba-sale'*) suddenly becoming scarce after former profusion. His later (1871) list reports it as common at the gravel pits on Coopersale Common. W.J. Lucas (1900a & 1900b) cited Harwood on *sanguineum* breeding on 'a part of the Essex coast' and as occurring at Colchester, and gave C.A. Briggs as the source of another report from Leigh. Harwood (1903) noted that *sanguineum* had a more restricted range than *striolatum*, mentioning St. Osyth as a locality, along with Epping Forest.

The Campion brothers appear to have seen this species only sporadically in the forest during the first decade of the 20th century, but did give evidence of its breeding in the Forest in 1908. They also reported the sighting of an immature specimen by A. Luvoni at Westcliff in 1912 (Campion & Campion 1913), and there are specimens in the Hope department of Entomology at Oxford collected by J.W. Yerbury at Frinton also in 1912. Large numbers were observed ovipositing at Langenhoe Wick, in the north-east in 1926 (Roberts, cited in Lucas 1927). Hammond (diaries), Syms (1929) and Pinniger (1933) report sightings in Epping Forest in the 1920s and 1930s, but Pinniger notes that it was seen rarely though regularly. It appeared in Hammond's unpublished notes in 1945, and by 1949 Longfield was able to report it as common on the Epping Forest ponds, Coopersale Common, Ongar Park and Hainault Forest. It was reported from Hainault Forest in 1934, Benfleet in 1936, and 1937 (Pinniger 1934, 1936, 1937 & Hammond, diaries, 1937).

D.A. Ashwell took specimens at Hatfield Forest in 1940 and 1941 (specimens in the London Natural History Museum) and Queen Elizabeth's Hunting Lodge museum has another taken by Bernard Ward at Hatfield Forest in 1949. Another specimen taken in 1947 by E.S. Brown at Hadleigh is in the Hope Department collection. Pinniger, Syms & Ward (1950) add Childerditch Common and Shelley.

Sympetrum striolatum (Charpentier, 1840) – Common Darter

The common darter is one of the most frequently encountered species during the autumn when, unlike most other species of darter, it regularly perches on the ground, or on reflective surfaces during cooler weather.

Description

The mature male common darter has an orange-red, narrow, parallel-sided abdomen, with small black markings on the dorsal surface of segments 8 and 9. Immature males and females are yellow to dull ochre, with females often taking on a reddish hue with age. Females have a prominent vulvar scale projecting below segment 8 (view from the side). The legs have a narrow yellowish longitudinal stripe in both sexes, and the black band across the forehead is not continued downwards at the sides.

Similar species

The most likely confusion is with the ruddy darter (*S. sanguineum*), which is smaller and more brightly coloured in the male. See under that species for detail on identifying features.

There could be confusion with uncommon migrant species of *Sympetrum*, but the combination of the shape of the black band across the forehead, ab-

*Male
common
darter*

Female
common
darter

domen shape, yellow line along the legs and lack of yellow tinting of the wings or colouration of the wing-veins are distinguishing features. For more detail see under *S. fonscolombii.*

Flight period

This species has one of the longest flight periods, with records obtained between dates of 23rd June and 29th November. The flight period during the current survey appears to begin some two weeks or more earlier than during the 1980s, but now, as then, this species, together with *Aeshna mixta*, is one of the latest to be seen in the Autumn. Small numbers survive into November in most years.

Habitat and behaviour

A restless species in favourable conditions, the common darter can be seen basking on fence posts, walls and other prominent positions from where it seeks its prey, frequently returning to the same perch. During the present survey, it was recorded from a wide variety of water systems such as canals and rivers, borrow dykes, gravel pits, ornamental lakes and garden ponds.

There were also many reports of individuals some distance from water, in woodland areas and on open ground. These are often teneral specimens as the common darter tends, like many other dragonflies, to disperse from

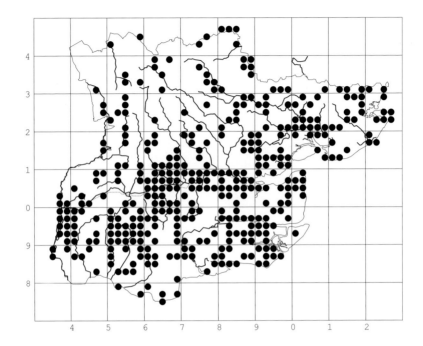

breeding sites after emergence. When they are fully mature the males take up territories in potential breeding sites, whereas the females are more widely dispersed. At this stage the males defend their territories aggressively, seeing off rivals of their own species as well as larger dragonflies, such as *Aeshna mixta*. Although females may sometimes be observed ovipositing solo or with the male hovering close by, typically the pair will fly in tandem in an undulating manner over the water or amongst emergent vegetation with the female releasing eggs as the tip of her abdomen is repeatedly flicked into the water, or onto vegetation.

The larvae take approximately a year to complete their development.

On sunny days in Autumn they bask on pale, reflective surfaces such as walls, fences, stones or even clothing.

Distribution and conservation

The common darter is widely distributed throughout Britain and Europe although in north-west Scotland and western Ireland it is replaced by the

Highland darter (*S. nigrescens*). This is sometimes considered to be a distinct species, but may well prove to be a darkened form of the common darter.

It seems likely that some of the common darters seen in Essex are in fact migrants from the continent. On 9th and 10th August 1998 increased numbers were observed at Bradwell-on-Sea (together with an influx of *A. mixta*) and ten were subsequently caught in UV moth traps between the latter date and the 30th September (Parr 1999). Similarly, nineteen were attracted to moth traps at Bradwell between 17th July and 9th October 1999 (Parr 2000).

The previous survey recorded the common darter from fifty-four of the fifty-seven 10km squares in the county, noting that 'it appears to be more local in the north-west of the county'. This was also confirmed by the current survey, during which this common species was recorded from fifty-five of the county's fifty-seven 10km squares.

Early records

The status and distribution of this species has changed little since Victorian times. Henry Doubleday (1871) described it as 'very common everywhere around Epping' and F.A. Walker (1897) recorded it (as *S. vulgatum*) on his list for Wanstead Park. Harwood (1903) wrote that it was 'abundant in many places about Colchester' and probably occurred 'freely throughout the county'. It was reported in lists for Epping Forest compiled by F.W. and H. Campion (1903–13), E.E. Syms (1929), C.O. Hammond (unpublished 1923–45) and E.B. Pinniger (1933). It was later recorded from Benfleet (Pinniger 1937; Hammond, no date) and Hatfield Forest (D.A. Ashwell, specimens in British Museum (Natural History) collected in 1940 and 1941; B.T. Ward, a specimen in Queen Elizabeth's Hunting Lodge Museum, dated 1949).

Sympetrum flaveolum (Linnaeus, 1758) – Yellow-winged Darter

Description

This is a medium-sized darter dragonfly with an orange-red abdomen in mature males, and yellow-ochre abdomen in females. The males also have black sides to the abdomen. The legs are black with longitudinal pale lines and the black bar on the forehead extends down the front edges of the eyes. Their most distinctive feature is the yellow-orange suffusion of large areas at the base of the hind wings, and also to a lesser extent on the forewings which is found on mature specimens of both sexes.

Similar species

The extensive yellow suffusion on the wings distinguishes this from the other *Sympetrum* species. In addition, the more parallel-sided, less vividly coloured abdomen and the pale lines on the legs distinguish males of this species from those of *sanguineum*. In *S. striolatum* the black band on the forehead does not extend down the edge of the eyes.

Flight period

On the European mainland this is reported to be from May to October. In Britain, migrant specimens are usually seen from late July to September, and locally bred ones from June.

Yellow-winged darter

Habitat and behaviour

It is reputed to oviposit in drier places than other *Sympetrum* species, including damp mud. Breeding sites are typically shallow, warm and well-vegetated ponds. There have been some successful breeding attempts in Britain, notably following the large immigration of 1995, but colonies have so far died out after a few years (Beynon 1998).

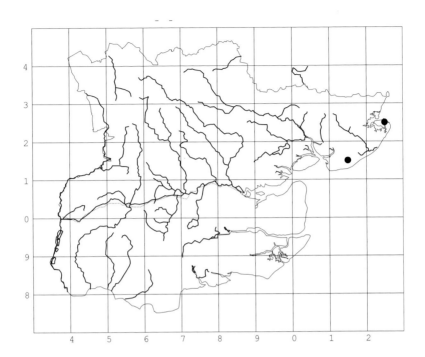

Status and distribution

The European distribution is mainly central, northern and eastern, with more scattered and often temporary populations to the south and west. It is an irregular migrant to Britain, but occasionally occurs in large numbers. 1871, 1898, 1906, 1926, 1945, 1955 and 1995 were notable 'invasion' years. There was another significant invasion in 1999 with reports of some 60 individuals across the country. There were only two reports in 2000 and none in 2002 or 2003. There was a small east coast influx in 2004 and again in 2005. At the time of writing it seems that 2006 will have seen a rather larger influx, again mainly on the east coast (Pittman 1996, Silsby 1995, Parr 1996, Silsby & Warne-Smith 1997, Parr 1997, Parr 1998, 1999, 2000, 2001, 2003, 2004, 2005, 2006).

Occurrence in Essex

Historical

On 16th September 1841 Henry Doubleday wrote to the *Entomologist* magazine to express his puzzlement at the appearance of this species 'in profusion' some five years previously but subsequent disappearance. He later (1871) reported it as 'very common' in certain years among gravel pits on Coopersale Common. It was abundant in the London district in 1871 (Lucas 1900b), and Stephens took it at Epping before 1837 (Longfield 1949). Several sightings were reported in 1898 (Lucas 1899), including one near Colchester seen by W.H. Harwood. The following year one was taken by Bernard Harwood at St. Osyth (Lucas 1900a).

Until 1995, subsequent records relate to Epping Forest. The Campion brothers observed a total of 14 males and one female in the Forest between 12th and 19th August 1906 (Campion & Campion 1906c). Specimens dated 12th and 19th August are in the Hope Department at Oxford, and another, dated 28/08/1908, is in the J.J.F.X. King collection (O'Farrell 1950). It is possible that this specimen represents a brief period of breeding success in the Forest following the invasion of 1906. 1926 was another invasion year, and again the species appeared in Epping Forest. Both males and females were reported as abundant in the Forest between 24th July and 23rd August (Lucas 1927a & b), with ovipositing noted between 8th and 11th August. C.O. Hammond's diaries include several reports for 1926, with Baldwin's Hill pond mentioned as a locality within the Forest. Another specimen was taken in the Forest by Hammond in August 1947 and exhibited at a meeting of the RES (C. Longfield, BRC).

We have not managed to locate any further records until the period (1980s) of the previous survey, when there were two reports, both from Epping Forest: One male was seen at Baldwin's Hill Pond on 19/08/1982 by C. Dale (EFCC records) and the other, a female, was seen by E.P. Ryan at Wake Valley Pond on 28/08/1984.

The next records we know of refer to the invasion year of 1995. Several species of *Sympetrum* arrived, with particularly large numbers of *S. flaveolum* landing along the east coast in early August with as many as 600 reported as congregating in a cemetery at Yarmouth. Essex saw its share of this immigration, with the first report (of a single male) coming from Woodham Fenn (EWT – TQ7997 (J. Hurley)) on 23rd July (also reported by A. McGeeney from the same site on 24th). As this site is more than 20km inland (although on the Crouch), it provided evidence for the view of Jill Silsby (pers. corr.) that there had been a smaller influx in the weeks prior to the main arrivals. A. McGeeney also saw one at Waltham Abbey on 24th July with 3 males seen later (6th August) at the same site. Another was seen at Fingringhoe Wick (EWT) during the first week in August by A. Goodey, following which small numbers (2–4) were seen at Old Hall Marsh (RSPB - TL963122) on 4th, 5th and 6th August (P. Charlton & I. Hawkins). Meanwhile approximately 18 were seen at Colne Point (EWT – TM1012) on 6th August (N. Harvey), and 4–5 seen at the same site by N. Cuming on 9th August. Further inland, 3 males were seen at the Backwarden reserve (EWT – TL7803) near Danbury on 6th August (P. & K. Gash), and also 2 at the same site on the 9th. Through August there were further sightings in the Danbury area: one in Pheasant House Wood and another in Poors Piece (EWT) in mid-August (P. Palmer), two at Hitchcock's Meadow (EWT) also in mid-August (J. Howchin), and Little Baddow Heath (EWT) on 16th August (the late G. Pyman). The latest record for the year also came from the Danbury area – 3rd September at the Backwarden (P. & K. Gash).

There was a scattering of reports from other parts of the county over the same period: B. Watts saw one on 6th August at Willowfield Farm irrigation reservoir, Tiptree (TL983148) and another at Fingringhoe Wick (EWT – TM047195) on 12th August. G. Smith saw one at Sandbeach Farm, on the Dengie Peninsula (TM024054) on 12th August, and a single male at Sales Point, Bradwell (TM029086) on the 13th. Another was seen by A. McGeeney on 28th August at Epping. One was seen at Rainham by J. Phillips on 28th August, and others were seen at Walthamstow on 10th August and 6th September (reported in Silsby & Warne-Smith 1997)

In 1996 there was considerable interest in the possibility of successful breeding, especially in the Danbury area, where individuals had been observed for several weeks in 1995. However, the only indication of possible breeding success was a record of an apparently freshly emerged female at Fryerning Hall fishing lakes, Ingatestone (TL633003) on 3rd June 1996 from G. Smith. One was seen in Riddles Wood (TM1217) on 29th July 1997 by R. Arthur and R.J. Seago.

During the present survey

The influx on the east coast between 7th August and 4th September 2004 (A. Parr 2005) included one Essex sighting: The Naze, Walton (TM265236) on 10/08/04 (C. Balchin & J. Rowland). We have no Essex records for 2005, but an influx of migrants from late July 2006 resulted in three Essex reports: at Colne Point (TM1012) on 13/08/06 (S. Cox); at Rainham Marshes, on 16/08/06 (unknown recorder, via A. Parr), and at Bedfords Park Lake (TQ518918), seen by C. Jupp.

Sympetrum fonscolombii (Sélys 1840) – Red-veined Darter

Description

The abdomen is red in mature males, and the front of the head also has a red area. The leading veins of the forewings in their basal half are also red. In most specimens the legs are black with a pale longitudinal stripe. In females the abdomen is yellow-brown, and the leading veins of the forewings are yellow in their basal half. In both sexes there is a small yellow suffusion at the base of the hind wings. The pterostigma is pale yellow-brown or reddish, with black borders.

Similar species

S. sanguineum: This has black legs (usually black with pale longitudinal stripe in *fonscolombii*); there is a marked constriction of the male abdomen (more parallel-sided in *fonscolombii*); the abdomen is relatively short and 'stubby' (relatively long and narrow in *fonscolombii*); the wing-veins are black (some are red or yellow in *fonscolombii*); and the yellow suffusion at the base is very small in extent (a little larger in *fonscolombii*).

Male red-veined darter (A. Kettle)

Red-veined darter – teneral specimen (A. Kettle)

S. striolatum: This has black wing-veins in both sexes (some yellow or red in the basal half of the forewings); the male abdomen is orange-red (brighter and more evenly coloured in *fonscolombii*); the yellow suffusion at the base of the hind wings is very small or absent (present, and slightly larger in *fonscolombii*); the black line across the forehead is not extended down the sides of the eyes (extended in *fonscolombii*).

S. flaveolum: Both sexes of this species have a large part (up to one third) of the basal area of the hindwings suffused yellow, and usually yellow areas on the forewings, too.

Flight period

In southern Europe the adults can be seen all year, though they are much less frequent in the winter. The northward migrations take place from late May onwards. Waves of migrant individuals usually arrive in Britain in June or July, but locally bred specimens are seen from late May, and there is sometimes a small emergence in late summer and autumn (from the end of August onwards). A few reports are from as late as early October.

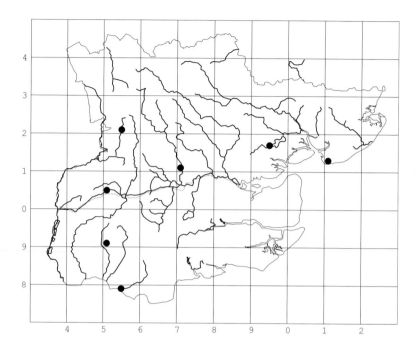

Habitat and behaviour

In southern Europe the species inhabits still-water sites, often small, shallow
ponds in quarries or coastal dunes. This is also true of favoured breeding
sites in Britain, where newly created farm reservoirs or flooded gravel pits
with areas of bare ground are characteristic sites. The males are territorial
and settle on favourite perches, as do other *Sympetrum* species. However,
they also spend much of their time sunning themselves on patches of bare
ground or stones. This is not noted in accounts of their behaviour in southern Europe and could possibly be a response to our cooler climate. They are
noticeably more irritable than other *Sympetrum* species, and usually difficult
to approach for photography. The life-cycle is short, with eggs laid in spring
or early summer producing adults later that year.

Status and distribution

The species is quite common in southern Europe, with more scattered populations further north, reinforced by recurrent migration years. Northward
migration, with evidence of attempted breeding has become more common

in recent years. There are records each year in Britain, with large migrations reported in 1992, 1996, 1998, 2000, 2002, 2003 and 2006. Numerous breeding attempts have been observed at sites in Britain, particularly in the south-west, and some colonies have survived for 3 to 5 years before dying out (see, for example, McGeeney 1997).

Occurrence in Essex

One male was seen at the Epping Forest Conservation Centre pond (TL413981) on 15/08/89 by E. Rothney, and another at Wake Valley pond in the Forest on the same date by C. Johnson.

During the period of our current survey, the species has been observed in 2002, 2003, 2004 and 2006. With the exception of our records for 2004 these correspond to 'invasion' years, with numbers of this species reported from numerous other UK localities. Evidence of breeding at Abberton Reservoir is significant, and suggests the possibility that this species may begin to colonise the county in future years.

2002

One seen near Epping by A. McGeeney on 19/06

One seen by M. Heywood on 19/06 (TL512056) at Ashlyn's farm, Bobbingworth. (conf. A. McGeeney)

One at Bradwell-on-Sea by S. Dewick, on 28/07.

2003

Single specimens seen at Bedfords Park lake (TQ518918) on 6th, 8th 11th and 14th June (possibly the same individual) (A. Middleton).

2004

One at Bedfords Park lake on 14th June (C. Jupp)

One male settled on a clothes line in a Colchester garden (photograph) (TM0025) between 16th and 20th August (S. Nixon).

2006

At a newly created pond, at the western end of Abberton Reservoir: on 17/06 two males, and on 1/07, as many as 20, including 3 pairs ovipositing. Also 2 males and a mating pair observed from Layer Breton causeway on the reservoir (A. Kettle). On 2/07 two males seen by G. Ekins on the reservoir. A. Kettle noted 15 males and one female at the original site on 29/07, and then, on 27/08 observed 5 teneral adults, indicating successful breeding at the site earlier in the summer.

At Bedfords Park lake C. Jupp saw up to 3 individuals from 18/06 to 21/06 and on 14th and 15th July.

Two were seen at Rainham Marshes by H. Vaughan on 28th June, two males at Channels Golf Club pools, Chelmsford on 1st July by G. Ekins and another at Writtle by T. Caroen on 28th July.

Considerable numbers were also seen by several observers at Stansted Airport lagoons (TL550217): up to 14 by S. Patmore *et al.* from 23/06 to 1/07. A male was reported from this site on 1/07 by G. Ekins, R. Cornhill & D. Acfield, and several more, settling on patches of bare ground, by R. Cornhill and T. Benton on 6/07.

Chapter 4
A history of dragonfly recording in Essex

Comparison of our present-day dragonfly fauna with that of earlier periods of history must inevitably rely to a considerable extent on guesswork. However, thanks to the dedicated and often innovative fieldwork of previous generations of naturalists in Essex our conjectures can at least claim to be *informed* guesswork. The development of the reliable recording of the distribution of any group of organisms is dependent on two important conditions. The first is the achievement of an accepted and stable nomenclature, and connected with this, the availability of reliable literature for use by field naturalists. The second condition is the growth of organised communication amongst naturalists themselves: the emergence of local and national societies concerned with natural history, the establishment of regular communica-

Henry Doubleday

tion between widely separated individuals and groups, and the existence of a periodical literature to stimulate, debate, adjudicate and give permanency to the results of field-recording.

Early studies of dragonflies

In the case of dragonflies, the first of these conditions was partially achieved by about the middle of the 19th century, but it was not until the turn of the present century that an adequate and reliable identification guide was available (W.J. Lucas 1900b). W.E. Leach was responsible for the first systematic classification of the British dragonflies (Leach 1815), separating out three families, the *Libellulida*, the *Aeschnides* and the *Agrionida*. The three genera *Gomphus*, *Cordulia* and *Cordulegaster* were also separated out from the hitherto very miscellaneous *Libellulida*. J.F. Stephens helped W.E. Leach in his work on the British Museum entomological collections and produced a list of British species of dragonflies. His *Systematic Catalogue of British Insects* (1829) did not contain descriptions, but his *Illustrations of British Entomology*, published from 1828 to 1846, included illustrations, some descriptions and notes on the distribution of dragonflies in vol. 6 (1835–7). Unfortunately, many of the names used by Stephens were misapplied, and he included a number of species now known not to occur in Britain. However, Essex localities figure among those given for several species, and these are among the earliest records we have for the county. Five species were given as occurring at Epping, significantly including *Aeshna mixta* and *Brachytron pratense* (under the name *Aeshna teretiuscula*). Epping was also given as a locality for *Cordulia aenea*, together with Woodford. *Lestes dryas* was said to be abundant in some of the 'marshy districts in the vicinity of the Thames', especially near Plaistow. This appears to be the earliest reference to *L. dryas* in Essex.

The study of dragonflies on the continent of Europe was far in advance of that in Britain. Charpentier (1825 and 1840) and Vander Linden (Linden 1825) were great pioneers, but the Belgian Baron E. de Selys-Longchamps was referred to by Tillyard (1917) as the 'father of Odonatology'. It was he who classified the dragonflies systematically on the basis of wing venation, and wrote monographs on all the sub-families of the Odonata (except the *Libellulinae*) between 1840 and his death in 1890. De-Selys Longchamps is particularly important for our story in that he visited Britain in 1845 to examine local collections and establish synonyms with continental nomenclature. The results of his findings were published in the *Annals and Magazine of Natural History* in 1846, and an abstract of his paper also appeared in the *Zoologist* (Selys-Longchamps 1846a, 1846b). Longchamps acknowledged the help of E. Newman and E. Doubleday and also mentioned the collections of Stephens, Leach and Curtis.

It was during this period, too, that national societies of entomologists were established, and with them a periodical literature which made possible a permanent record and wider dissemination of knowledge about orders of insects such as dragonflies. These societies tended to be rather élite associations of learned gentleman-naturalists, quite unlike the more popular local clubs and societies that became more widespread later in the century. They were nevertheless very important in providing the scientific groundwork and stimulus from which the later societies benefited. Two of the earlier societies – the Entomological Club and the Entomological Society of London (later the Royal Entomological Society) – were of particular significance to the history of Essex entomology. The eminent Quaker naturalist Edward Newman played a key role in both societies. The Entomological Club was founded in 1826 and established the *Entomological Magazine*, with Newman as its first editor, in 1832. By 1836, the club possessed reference collections and a library. Newman was also instrumental in the founding, in 1833, of the Entomological Society of London, and was one of its presidents.

The Doubledays

Amongst Newman's closest friends and collaborators were two Essex entomologists (and fellow Quakers) Edward and Henry Doubleday. The Doubleday brothers were born, and spent their early years together, at Epping, where their parents ran a hardware and grocery business. Robert Mays (1978), Henry's biographer, tells us of the wide-ranging interests and delight in the natural history of Epping Forest and its vicinity, that the brothers shared. Henry remained in Epping, taking over the family business when their father died in 1847. Edward, the younger brother, had by then left home, spending two years in the USA collecting specimens for the British Museum. He subsequently held a post in the Museum, working mainly on Lepidoptera until his early death at the age of 38 in 1849.

In Volume I of the *Entomological Magazine* (1832–30) Newman's regular 'Entomological Notes' include some discussion of the classification of the Dragonflies (*'Libellulites'*) and are notable for his view that Dr Leach's genus *Libellula* (then a very large and heterogeneous grouping) would eventually resolve into three genera, and for his listing of the species of one of these genera, *Sympetrum*, which bears his name to this day (Newman 1832–3). Newman also ventured to question Vander Linden as to the status of a type of dragonfly taken by 'my friends' H. and E. Doubleday at Epping. This was the *praenubila* of *Libellula quadrimaculata*. Newman mistakenly regarded it as a separate species.

Volume III of the *Entomological Magazine* contains the first significant local list of Essex Odonata. This is contained in a letter from Edward Doubleday (E. Doubleday 1835), dated 21st May 1835. In the letter, Doubleday describes a locality to the east of Epping adjoining the woods (including Ongar Park

Woods) owned by Capel Cure of Blake-hall. Here there is a small portion of forest, with 'many open boggy places' and vast numbers of gravel pits. Doubleday's list includes the following dragonflies: *Anax formosa* (= *Anax imperator*), *Aeshna teretiuscula* (= *Brachytron pratense*), *Gomphus vulgatissimus*, *Cordulia aenea*, *Libellula quadrimaculata*, and *Agrion rubellum* (= *Ceriagrion tenellum*). Doubleday also lists *Libellula praenubila* as a separate species, carefully expounding his reasons for considering it such. *C. tenellum* is an acid-bog species, for which there are no 20th Century Essex records, so that it is noteworthy that Doubleday reports it as occurring 'in profusion'. Referring to *C. aenea*, Doubleday writes of the 'hundreds which swarmed over these pits'. Unfortunately, Doubleday gives us no further details about the status of *G. vulgatissimus* whose presence near Epping even in those days is somewhat surprising.

The *Entomological Magazine* was soon discontinued, to be replaced, also under Newman's editorship, by the *Entomologist*, which first appeared in 1840. This magazine, too, was very short-lived, but did carry two reports on Essex dragonflies. The first of these was a report by Edward Doubleday (E. Doubleday 1841) on the capture by one Marsh of a 'single specimen of the rare dragonfly *Sympetrum rubicundum*', on 1st June 1841. This is probably the species now known as *Leucorrhinia dubia*, yet another species of acid-bog dragonfly not recorded in Essex in the 20th century. This is followed by a note from Henry Doubleday dated 16th September 1841 (H. Doubleday 1841) in which he discerns 'something strange in the sudden appearance and disappearances of certain *Sympetrum* species'. *S. flaveolatum* (= *S. flaveolum*) and *S. basale* (probably = *S. sanguineum*) appeared in profusion some years previously, but since then either disappeared altogether (the former species) or became scarce (the latter). The appearance of *S. rubicundum* (presumably the one referred to by Edward, though the date is now given as May) and *S. scoticum* (= *S. danae*), the latter in profusion, in 1841, where neither had been seen in previous years is also noted. Certainly in the case of *S. flaveolum*, and probably in the case of *S. sanguineum*, Doubleday's puzzle may be solved by our present-day understanding of insect migration. *S. flaveolum* still is seen occasionally in Essex as a migrant, but there is no evidence of it as a breeding species in the county. As for *S. danae*, Doubleday may have witnessed the first establishment of this species in the Epping Forest area (and probably in Essex as a whole, since the species has only rarely been recorded in Essex, outside the Forest).

In 1843 it was decided to merge the *Entomologist* and the *Zoologist* with Newman, again, as editor (a role he retained until his death thirty years later). A second series of the *Entomologist* was launched in 1864, and this rapidly became an indispensable means of communicating entomological opinions and reports, especially local species lists. With the death of Edward Doubleday and Henry's preoccupation with his major work on the Lepidoptera, there is a gap of some thirty years in the recording of dragonflies in Essex. During that time, however, great strides were made in the scientific study and classification of Odonata. Some ten years after the appearance of Longchamps' corrected list and synonyms for the British species, his close collaborator, Dr H. Hagen, published 'A Synopsis of the British Dragonflies' in *The Entomologist's Annual* (Hagen 1857). This synopsis includes those continental species most likely to be discovered in Britain, and adopts a system of classification and nomenclature close to that now in use, with the exception that Newman's proposals on the genus *Libellula* are not followed, so that all the 'darter' dragonflies continue to be 'lumped' together in this one genus.

The next major publication on the British Odonata was *A Catalogue of British Neuroptera* (McLachlan & Eaton 1870), which appeared under the auspices of the Entomological Society of London. McLachlan was a friend of Longchamps and a noted specialist in Trichoptera and Neuroptera. In his catalogue, the Odonata comprise six families in the sub-order Pseudo-Neuroptera and are classified after Longchamps and Hagen, with the difference that Newman's sub-division of the genus *Libellula* is at last adopted. McLachlan lists forty-one species as British, omitting doubtful records and 'casuals' not recorded for many years.

When Henry Doubleday eventually returned, in print, to the topic of dragonflies (H. Doubleday 1871) he began with a complimentary remark on McLachlan's recently published 'valuable catalogue'. What follows is a list of species claimed by Doubleday as occurring in the neighbourhood of Epping, but with the proviso that 'some of the best localities are destroyed, and I am not certain that all the species enumerated are now to be found here' (p.86).

Doubleday's astonishing list of no less than thirty species cannot, unfortunately, be taken at face value. He was a renowned ornithologist, and a Lepidopterist of major international status, but his work on dragonflies seems to have been a rather more casual and intermittent affair. It is also worth noting that at the time he presumably compiled this list (July 1871) he had only recently returned from the Quaker mental hospital in York, having previously suffered from a serious breakdown in his health. He suffered continuing financial and health problems, and eventually his belongings were put to auction in 1871. He died in June 1875 (see Mays 1978). Despite Doubleday's great reputation, several writers who continued to refer to his important list nevertheless expressed scepticism about some of his claims. *Calopteryx vesta* is merely a synonym of *Calopteryx virgo*, but Doubleday insisted on the presence of a distinct species under this name. In a note added to Doubleday's published list, McLachlan himself cast doubt both on this, and on Doubleday's report of *Lestes virens*. McLachlan's views were repeated at greater length in the *Entomologists' Monthly Magazine* (1884). Later on W.J. Lucas (1900b) was also sceptical about Doubleday's report of *Lestes dryas*, though what is now known about the habitat preferences of this species suggests that the record cannot entirely be ruled out.

The remaining list of twenty-seven species still contains some remarkable records by modern standards. *Gomphus vulgatissimus* and *Libellula fulva* are both rare moving-water species, which, in the case of the former species, appear not to have been recorded in Essex since Doubleday's list. Edward Pinniger recalls having discussed Doubleday's records with W.J. Lucas (then an old man). At that time the record of *G. vulgatissimus* 'seemed highly unlikely'. Since then Pinniger's experience of the liability of this species to wander far from its river of origin has led him to consider that 'perhaps Doubleday was right after all, alas we shall never know' (pers. comm., 2.9.1985). *C. tenellum* and *L. dubia* are species of acid-bog and wet heathland for which, again, we have no Essex records subsequent to Doubleday's. However, they do occur on earlier lists – E. Doubleday (1835) in the case of the former species, Evans (1845) in the case of the latter. In view of this and the major habitat changes which have occurred in the Forest (some of them already remarked upon by Doubleday himself) there seems to be no particular reason to doubt that these species did formerly occur there. *Ischnura pumilio* has seemed to be a very doubtful record, but the unexpected discovery of populations of this species in chalk quarries in Bedfordshire and Buckinghamshire in the late 1980s renders it a little more believable. Further information on these and other early records is given in **Appendix 1**. Another unlikely inclusion in Doubleday's list is his report of *Aeshna mixta* as rare, being found on North Weald Common in June. As pointed out by the Campion brothers (1905a) the early date rules out this species: *A. mixta* is rarely on the wing before late July. Apart from Stephens (1835–7) there appear to be no reliable Essex records for *mixta* until close to the end of the 19th century.

Towards the first county list

After Doubleday's list, there is a gap in our knowledge of Essex dragonflies of some twenty years. During this time Epping Forest was saved from illegal enclosures by local landowners, and secured for the public (see R.M. Sharp 2007, forthcoming). However, as we shall see, this was not sufficient to ensure adequate preservation of the wildlife including the dragonflies of the Forest.

In the early 1890s there was a renewal of interest in dragonflies and numerous local lists were published. It seems probable that this was stimulated by the publication, in 1890, of W.F. Kirby's *Synonymic Catalogue of Neuroptera-Odonata, or Dragonflies* (Kirby 1890b). Kirby built on the work of Longchamps but incorporated his own earlier revision of the sub-family *Libellulinae* (Kirby 1890a). Whilst of less scientific value, W. Harcourt Bath's *Illustrated Handbook of British Dragonflies* (Bath 1890) was probably more effective in popularising the study of dragonflies. Harcourt Bath also intervened in the pages of the *Entomologist* (Bath 1893) in response to a local list of dragonflies of the Chester district, bemoaning the lack of entomologists interested in dragonflies, and calling for more local lists to provide material for a forthcoming publication on the distribution of dragonflies. Indeed, the relative lack of popular interest in such eye-catching and fascinating insects as dragonflies, compared with, for example, butterflies and moths, is rather surprising. One clue to the answer to this puzzle is the frequent discussion in the 19th century literature of techniques for preserving the colours of set specimens. The tendency for the magnificent colours of many of our dragonflies to fade after death made them comparatively poor subjects for the collecting mania which gripped the Victorian gentleman-naturalists. This fact was undoubtedly very beneficial to dragonflies, but probably delayed the growth of our knowledge of them.

Harcourt Bath's popularising enthusiasm for the dragonflies further expressed itself in an article on 'Some Famous Collecting Grounds for Dragonflies' which extended over two issues of *Science Gossip* (Bath 1892). Here he extolled 'the delightful domain of Epping Forest' which ranked 'second to none in England for the richness of its Dragonfly fauna'. As well as listing several of Doubleday's specialities, Harcourt Bath also mentioned the occurrence of *Agrion* (= *Coenagrion*) *mercuriale*, supposedly taken by a Mr W.H. Nunney. We can find no other reference to this discovery in the literature, and in view of what is now known of the habitat requirements of *C. mercu-*

riale, it seems unlikely that the species ever did occur in the Epping area. However, it is significant that Harcourt Bath's article is extensively quoted in the *Essex Naturalist* (vol. vi, 1892, pp.44–45), since it is from about this date that entomologists in other parts of Essex seem to have taken a greater interest in dragonfly recording.

In 1894, possibly under the influence of Harcourt Bath's appeal in the previous year, the *Entomologist* published the first of what was to be a series of annual reports on dragonflies written by W.J. Lucas. The reports are a mixture of collated communications from observers in various parts of the country, together with first-hand accounts of Lucas's own observations, including descriptions of the immature stages of several species. Local lists from now-famous sites in Surrey and Hampshire predominate, but several reports do concern Essex sites. The Rev. F.A. Walker of Cricklewood contributed a short piece to the *Entomologist* of 1897 (Walker 1897) on the 'Dragonflies of North London' in which he gave Wanstead Park as the best local site, listing twelve species as occurring there.

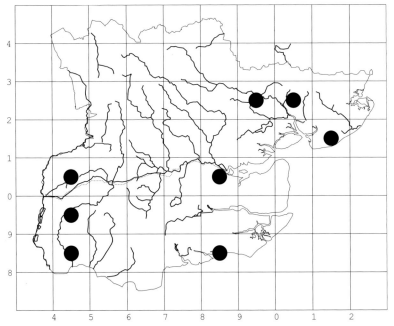

Dragonfly recording in Essex – coverage prior to 1903

Lucas's annual reports for 1898 and 1899 made extensive reference to sightings made by the eminent Colchester entomologist, W.H. Harwood. The rare migrant dragonfly *Sympetrum flaveolum* was reported from Colchester in 1898, along with others from Surrey and Oxford. Lucas's report for 1899 quoted Harwood explicitly as the source of another specimen of *S. flaveolum* taken at St Osyth. Since this coastal locality seemed an unlikely breeding site, Lucas saw this as confirmation of his view that this was an immigrant to Britain. Harwood was also reported as having seen *A. mixta* 'disporting' over the River Colne in Essex a 'few days before Oct. 21st'. The lateness of the date supports Harwood's identification and makes this the first acceptable record of this species in Essex since Stephens (1835–7). Finally, *Sympetrum sanguineum* was reported by Harwood as breeding on part of the Essex coast. It still does so today, but has recently become much more widespread.

W.H. Harwood was born in Colchester in 1840, and lived there, at various addresses, until 1915, shortly before his death in 1917. For most of his adult life Harwood was a professional entomologist, and dealer in natural history specimens. He is best known for his work on the Lepidoptera (see Firmin *et al.*, 1975, pp.11–12, and Firmin 2003, pp.12–15), but he and his two sons, Philip and Bernard S., developed a close and detailed interest in other insect orders and their north-east Essex habitats. W.H. Harwood maintained a correspondence with many of the leading entomologists of the day, including Edward Newman and W.J. Lucas. He contributed regular notes and articles to the entomological literature, and was actively involve in the work of the Essex Field Club from its early days. Harwood senior developed his interest in the more neglected orders of insects later on in life, taking up the Coleoptera and Hymenoptera (Aculeata) in the early 1880s, and other orders, such as the Odonata still later. Volume III of the *Essex Naturalist* (1889) reports his presentation of a paper to the Field Club on the rarer Coleoptera of the Colchester district, together with an accolade as 'a name known all over England' from E.A. Fitch, then the President of the Club. Harwood's sons seemed to have shared his enthusiasm for the 'other orders' during these closing decades of the century, judging from their surviving diaries (kept, now, by the Hope Department of Entomology, Oxford University Museum), and by their contributions to the entomological literature of the time.

Naturally enough, it is to Harwood that we are indebted for the first genuinely county-wide report on the dragonflies of Essex. This is given in his section on 'Insects' in the Victoria County History (Harwood 1903). Harwood listed twenty-eight species, but seven of these (*L. dubia, S. scoticum* (= *S. danae*), *L.*

fulva, G. vulgatissimus, C. tenellum, I. pumilio and *C. pulchellum*) were merely drawn from Doubleday's (1871) list without further information. Another of Doubleday's species, *L. virens* was bracketed as doubtfully British. Harwood also incorporated F.A. Walker's Wanstead list, with the possibly significant exception of the latter's report of *Coenagrion pulchellum*. Harwood's own records of *Sympetrum flaveolum* were also included, along with notes on a 'swarm' of another migratory species, *Libellula quadrimaculata*. Interestingly, *Anax imperator* was given as 'now a rarity in the county'. *Aeshna mixta* was still considered to be a rare and local insect but was extending its range and had increased its numbers over the previous two years.

Harwood credited one 'Mr C.R. Briggs' with two particularly interesting records. One is for *Platycnemis pennipes*, a subsequent record to its appearance on Doubleday's list, but Harwood gave no further details. The other record from Briggs was for *Lestes dryas*, 'one of our rarest species', a single specimen of which was taken by Briggs near Leigh in 1891. Harwood's source was almost certainly the entomologist C.A. Briggs, a Fellow of the Entomological Society of London and a member of the South London Entomological and Natural History Society (later the British Entomological and Natural History Society) in the closing decades of the century. Briggs was primarily a lepidopterist, but acquired an interest in Odonata and Ephemeroptera from about 1892.

Briggs was also quoted in W.J. Lucas's classic book *British Dragonflies* (1900b). His records of *L. dryas* and *P. pennipes* were mentioned by Lucas, as well as a record of *Sympetrum sanguineum* from Leigh. Lucas also included a record of *Libellula depressa* from the Maldon naturalist E.A. Fitch. One of the founders of the Essex Field Club, and its president for ten consecutive years, Fitch was particularly interested in the Hymenoptera and other neglected orders of insects. The joint efforts of these field naturalists added to Doubleday's Epping list records from a range of Essex sites including Colchester, Wivenhoe, St Osyth, Clacton, Birch, Maldon, Leigh, Wanstead and Woodford, but significantly, added no new species to Doubleday's list. This indicates a preeminence of the dragonfly fauna in the vicinity of Epping *vis à vis* the rest of the county which persists to this day.

From 1900: the focus on Epping

After the publication of Lucas's splendid work, reports of field observations of dragonflies became much more frequent, though, as far as Essex is concerned, Epping Forest remained the focus of attention for several decades. The Doubleday's worthy successors as recorders in the Forest were F.W. and H. Campion, of Walthamstow. Their yearly reports on the Forest dragonflies appeared in the *Entomologist* between 1903 and 1909 (see Table 5), and there are subsequent articles under joint authorship and by H. Campion alone on specific aspects of dragonfly natural history, such as their prey and parasites. The Odonatist R.J. Tillyard in his standard work *The Biology of Dragonflies* (1917) acknowledges Herbert Campion's help, and cites several of their published articles. Though they were not able to confirm the continued existence in the Forest of many of Doubleday's rarities, the Campions did manage to add two new species to the county list. The first was their capture of a single male of the species *Orthetrum coerulescens* near Chingford on 22nd July 1900 (Campion & Campion 1906b). It seems unlikely that this represented a breeding colony, but there are a few subsequent records in the county.

	1902	1903	1904	1905	1906	1907	1908	1909
C. splendens		+						(+)
L. sponsa		+	+		+	+	+	(+)
P. nymphula	+	+	+	+	+	+	+	+
I. elegans	+	+	+	+	+	+	+	+
C. puella	+	+	+	+	+	+	+	+
E. cyathigerum	(+)	+	+	+	+	+	+	+
E. najas							+	+
B. pratense					+	+		+
A. cyanea	+	+	+	+	+	+	+	+
A. grandis	+	+	+	+	+	+	+	+
A. mixta	+				+			
A. imperator	+			+	+	+	+	+
C. aenea					+	+	+	+
L. depressa	+		+	+	+	+	+	+
L. quadrimaculata	+				+		+	
S. danae					+			
S. sanguineum		+			+	+	+	+
S. flaveolum					+		+	
S. striolatum	+	+	+	+	+	+	+	+
S. vulgatum					+			

Table 5 Dragonflies recorded by F.W. & H. Campion in Epping Forest 1902–1909. Brackets refer to records outside the Forest boundary

The extremely valuable annual lists provided by the Campion brothers include twenty species (not counting *O. coerulescens*). Even taking into account the probable errors in Doubleday's (1871) list, the Campions' lists suggest a marked decline in the Forest Odonata prior to the turn of the last century. With one exception (*Sympetrum danae*) the species associated with acid bog had disappeared and several moving-water species mentioned by Doubleday either do not appear on the Campions' lists (*Platycnemis pennipes*, *Coenagrion pulchellum* and *Calopteryx virgo*) or do so very irregularly (*Calopteryx splendens*). This indicates that quite marked habitat changes had been occurring in the Forest over the previous half-century. Three species (*B.pratense*, *C. aenea* and *S. danae*) were not seen by them until 1906, and were noted somewhat intermittently thereafter. This suggests that these species were not common – or at least were rather localised – in the Forest at that time. *A. mixta* and *S. sanguineum* are other species apparently recorded rather irregularly. The Campions' lists are also notable for the inclusion of records of two rare migratory dragonflies – *Sympetrum flaveolum* (also recorded earlier by Henry Doubleday) and *Sympetrum vulgatum* (a new county record).

There is little surviving information on the dragonfly fauna of Essex outside Epping Forest during the early years of the 20th century. Specimens of two species, *Sympetrum sanguineum* and *Lestes sponsa*, from Frinton and Walton-on-the-Naze, respectively, were collected in 1912 by J.W. Yerbury. These are in the Hope Department Collections in Oxford. Also from this period, the Campions' dragonfly notes (Campion & Campion 1913) for the 1912 season mentioned a report from A. Luvoni of an immature male of *S. sanguineum* at Westcliff on 14th July of that year.

The inter-war years, and especially the period following the publication of Cynthia Longfield's superb *Dragonflies of the British Isles* in the 'Wayside and Woodland' series (Longfield 1937), saw a marked increase in the study of the Essex dragonfly fauna, with a circle of first rate field naturalists exchanging information and encouragement, and publishing their findings in both national and local journals. The *Essex Naturalist* carried many of these reports, as did the *London Naturalist*, the journal of the London Natural History Society, from the early 1920s onwards. This circle of naturalists included Cyril O. Hammond and Cynthia Longfield, two of the last century's great experts on the dragonflies, together with Edgar E. Syms, Hugh Main, Bernard T. Ward, Edward B. Pinniger, R.M. Payne, L. Parmenter, K.M. Guichard and others. Though our main concern is with Odonata recording, the contribution made by these naturalists to our knowledge of the ecology, botany, Orthoptera, Neuroptera, Diptera, Ornithology and other aspects of our county's natural history is immense. Their work deserves to be much more fully recognised in print.

Cyril O. Hammond was studying the dragonflies of Epping Forest from as early as 1923, and his extant diaries contain notes on his observations through to the mid-1940s. Hammond's notes suggest that the dragonfly fauna of the Forest was little changed since the Campions' studies and, like them, he recorded twenty species there. *C. splendens* reappears on the list whereas *A. mixta*, recorded only sporadically by earlier observers, does not appear to have been recorded in the Forest by Hammond until 1937. *Sympetrum flaveolum* appeared in the Forest in considerable numbers in July and August 1926 (two of Hammond's specimens, collected in that year, are in the W.J. Lucas collection, at the Hope Department, Oxford (see also Lucas 1927a and 1927b). Both *C. aenea* and *B. pratense* were recorded regularly through the 1920s whilst *C. aenea* continues to be included through to 1945. The other Epping Forest rarity, *Sympetrum danae*, was not found by Hammond until 1926, when he commented that it 'never became common'. *Orthetrum cancellatum* was first recorded for the Forest in 1937, in which year also *Sympetrum danae* was reported to be common.

Edgar E. Syms (1881 to 1966), who published a further list for Epping Forest in 1929, lived for many years at Wanstead. He was a handicrafts teacher in East London, and according to F.D. Buck (1966), in his younger days he was 'an ardent trade unionist, a soap-box orator and something of a firebrand'. Perhaps his political background explains Syms' extraordinarily active role in the Essex Field Club, the London Natural History Society and the South London Entomological and Natural History Society, not to mention his fellowship of the Royal Entomological Society, the Zoological Society of London and the Amateur Entomologists' Society.

Syms joined the Essex Field Club in 1913, and was its president from 1956 to 1959. He was curator of the Passmore Edwards Museum for several years, and was elected an Honorary Member of the Club in 1963, some three years before his death in 1966. Between 1929 and 1960, the pages of the *Essex Naturalist* are studded with references to exhibits and talks presented to the Field Club meetings on a wide range of entomological topics: Diptera, the breeding of crickets, earwigs, stoneflies, beetles, solitary bees, potter wasps, moths, and, of course, dragonflies. Syms was also a keen photographer, and there are several references to the use of his 'lantern photographs' at Field Club meetings.

In the LNHS, Syms' involvement seems to have been mainly with the Entomological section (formed in 1925), and there are reports of his having spoken to meetings of the section on British Neuroptera (1938), Plecoptera (1946) and the breeding of insects (1953). Syms joined the 'South London' in 1916, was Assistant Secretary from 1928–1930, indoor meetings secretary from 1938 to 1952, Hon. librarian for twenty-six years – from 1926 to 1952 – and was voted an Honorary Member in 1960. At his death, Syms' lantern slides were donated by his son to the South London (now 'British') Entomological and Natural History Society. Though Syms certainly was a collector, of books and periodicals as well as insect specimens, he had a great interest in the life-histories of insects, and was above all a field-naturalist. F.D. Buck tells us that Syms, like Hugh Main, a great friend of his, was 'more concerned with living insects than with cabinet specimens' (Buck 1966). Several of the younger naturalists who knew both Syms and Hugh Main have paid tribute to their kindness and generosity in helping and advising beginners.

Edgar Syms' first published work on the Essex dragonflies (Syms 1929), was, like so many previous lists, confined to Epping Forest. His article begins with a useful account of the classification, structure and reproduction of dragonflies, and goes on to present a list of the dragonflies of the 'neighbourhood of Epping Forest' for the previous fifteen years (i.e. covering the period between the last report of the Campions, and the late 1920s). The list contains twenty-one species and, in a separate note, *S. flaveolum* is reported as having been recorded for the Forest (presumably not by Syms). The three scarce Forest species, *S. danae*, *B. pratense* and *C. aenea* are included in both Syms' and Hammond's unpublished lists, and also, like Hammond, Syms appears not to have seen *A. mixta* in the Forest up to this date (1929). The appearance of more moving-water species such as *Calopteryx virgo* and *Coenagrion pulchellum* may indicate an improvement in the status of these species since the period of the Campions' lists, or may have to do with the ill-defined limit of the boundaries of the survey area (two of the best slow-moving water sites in Essex – the Rivers Lea and Roding – run close to the Forest).

One of the younger naturalists who was helped and encouraged by Hugh Main and Edgar Syms was Edward B. Pinniger. Although he spent the later years of his life in Berkshire, Pinniger's home was in Essex for some sixty-eight years, first at Higham's Park, and subsequently, from 1927, at Chingford. Then in his mid-teens, Pinniger was able to explore Epping Forest by bicycle. His interest in natural history was encouraged by his parents, and his attention was turned to dragonflies in particular by Hammond and also by his contemporary at school, J.D. Gillett, who also knew Hammond. Early on, Pinniger joined both the LNHS and the 'South London'. At meetings of the latter, he met W.J. Lucas, then an old man, and the great specialist on dragonfly life-histories, A.E. Gardner. By this time also, Cynthia Longfield was active in the LNHS with its newly founded Entomological Section. It was she, together with C.L. Collenette, who encouraged the young Pinniger to publish his research on the dragonflies of the Forest.

Pinniger's talk on the 'Paraneuroptera of Epping Forest' was delivered to the Entomological Section of the LNHS in 1932 and was published in the *London Naturalist* for that year (Pinniger 1933). The article gives useful references to the main earlier contributions to our knowledge of the Epping Forest Odonata, together with descriptions, flight periods and so on of all the recorded species. Though believing at least twenty-two species occurred in the Forest, Pinniger himself had found just twenty. He had been unable to detect either *Aeshna mixta* or *Brachytron pratense*, but reported published sightings of these species by other observers. On the other hand, *Coenagrion pulchellum* and *Calopteryx virgo*, absent from some earlier lists for the Forest, were reported by Pinniger. Another interesting re-appearance is *Platycnemis pennipes*, but the report is of a singleton only, found in 1932, and may not, therefore, represent a breeding population within the Forest.

Pinniger continued with annual reports on the Odonata of Epping Forest, sometimes with notes on other localities in Essex and acknowledging records from Longfield, L. Parmenter and K.M. Guichard (at that time secretary of the Entomological Section of the LNHS. See Else 2002), until 1937. The report for 1933 (Pinniger 1934a) gives a total of fourteen species on the wing in the Forest by mid-June. *B. pratense* was seen for the first time in the Forest by Pinniger in that year 'in some numbers', and *E. najas*, too, was now to be found on nearly every pond in the Forest. However, only one specimen of *S. danae* was seen, confirming his earlier comment on the liability of this species to fluctuate in numbers from year to year. The Rivers Lea and Roding were quoted as sites for *C. splendens*, but *C. virgo* was not seen. In the same volume of the *London Naturalist* (Pinniger 1934b) Pinniger discussed the puzzling status of *A. mixta* in the Forest. Despite several searches in 1933 he had 'failed to produce any definite evidence' about this species. It will be remembered that Hammond's unpublished notes contain no Forest record of *A. mixta* until 1937.

Pinniger's dragonfly notes for 1934 (Pinniger 1935) included reports from the marshes on the River Lea below Chingford, the River Roding (where *P. pennipes* was found), Hainault Forest, and 'the Essex Marshes', as well as a more detailed report on Epping Forest. The Hainault Forest report is notable for the discovery of *Orthetrum cancellatum*, as Pinniger suggested, probably the first record of this species in the county. *Lestes dryas* was said to still persist 'in a favoured spot on the Essex Marshes', while *B. pratense* was now reported as 'very common' in Epping Forest. At last Pinniger was able to report his own sighting of *A. mixta* on 8th September 1935 (Pinniger 1936) in the Forest. *Lestes dryas* was a subject of attention in the report for 1936. Formerly considered a great rarity, the species was found 'in considerable numbers' in the 'salt marshes of the River Thames', and was also seen at Southend and, 'in numbers' at Burnham in Essex. The report for 1937 (Pinniger 1938) gave *A. mixta* as common in the Forest, and there was a further report (via K.M. Guichard) of *O. cancellatum*, at Mill Hill. Benfleet was given as a locality for *L. dryas* (both Pinniger and Hammond gave 18th July as the date, but their lists for the day differ in other respects), as well as *S. sanguineum*, *A. imperator* and *L. depressa*.

In 1935 discussions took place which resulted in the formation of an Ecology Section of the London Natural History Society, and in 1936 Pinniger, now chairman of the Chingford branch of the Society, reported on the consideration by the branch of the possibility of future ecological work on the Forest. The idea took some time to be translated into practice, but the Chingford branch's survey of the Cuckoo Pits area of the Forest eventually started in 1942. The *London Naturalist* for 1942 indicated that the Odonata had been studied, but no details were given. However, draft notes by Pinniger dating from 1942 give details of fifteen species of Odonata observed from the survey area that year. These include one new record for the Forest – *Orthetrum cancellatum* – one female of which was seen on 27th June. Pinniger noted, in the version of the report eventually published in the London Naturalist for 1944 (Pinniger 1945), that the species may breed in the newly formed bomb craters (outside the survey area as strictly defined). Another species given as breeding in these craters was *Libellula depressa*. A letter from Cynthia Longfield to Pinniger (now in the Epping Forest Conservation Centre) dated 11th September 1943 also reported *Anax imperator* and *Sympetrum striolatum* from the bomb crater area. In her later (Longfield 1949) publication on the Dragonflies of the London area Longfield noted how 'exceedingly quickly' the dragonflies colonised these bomb craters. The Cuckoo Pits themselves were former gravel pits, but even during the three years from 1942 to 1944 with which the published report deals they had become noticeably drier, so that only one pond had significant standing water through the summer of 1944. Only eight of fifteen species were of confirmed breeding status in the Pits themselves.

The 1940s: Essex beyond Epping Forest

The publication of Cynthia Longfield's classic work *The Dragonflies of the British Isles* (Longfield 1937. See also the biography of Longfield, Hayter-Hames 1991) stimulated a much wider interest in the study of dragonflies, and this, together with the growth of local natural history societies gave rise to a steady expansion of reports on dragonflies in Essex outside the much favoured areas of Epping Forest and the London fringe. The Bishops Stortford naturalist, D.A. Ashwell, made a study of the dragonflies of Hatfield Forest between 1935 and 1945 (J. Fielding, pers. comm.), recording eighteen species during that period. Specimens of fifteen of these species are in the collections of the British Museum (Natural History) and the Chelmsford & Essex Museum. These all date from the years 1939 to 1942, and include two specimens (dated 9/6/1940 and 1/7/1940) of *Cordulia aenea* captured in Hatfield Forest. These are the only evidence of an Essex colony of this species outside Epping Forest during the last century. *Aeshna mixta, Orthetrum cancellatum* and *Sympetrum sanguineum* all appear on Ashwell's list for this period, as does *Lestes dryas* although it has not been possible to locate a surviving specimen of this species. Another remarkable record for Hatfield Forest is *Platycnemis pennipes*, seen in June 1939. This was, however, a singleton and probably not indicative of a breeding colony at that time. Ashwell, who was then secretary of the Bishops Stortford Natural History Society, led an excursion of the Essex Field Club to Hatfield Forest on 18th July 1948, but, unfortunately, the weather was poor and only three species were seen.

Ashwell was also responsible for a number of valuable records from other parts of Essex. There are specimens dating from between 1939 and 1957 from the Chelmer/Blackwater, just north of Little Baddow. These include *I. elegans* and *C. splendens*, as well as *P. pennipes* (1942 and 1957). Although it has not been possible to locate specimens, Ashwell is also credited with records of *B. pratense* from this site between 1940 and 1960 (R. Merritt, pers. comm.). Finally, there is a record of *C. virgo* from Plaistow, taken in June 1937.

Meanwhile, in the south of the county, interest had previously focussed on the search for *Lestes dryas*, (see Proceedings of the South London Entomological and Natural History Society, 1933–40). The newly formed South Essex Natural History Society should have been well-placed to carry out much-needed field work on the Odonata in south-east Essex. But H.C. Huggins, the well-known lepidopterist (see Firmin *et al.*, 1975) bemoaned the lack of entomologists concerned with other orders in his Entomological Notes for the January 1939 issue of the Society's *Bulletin* (Huggins 1939). Huggins claimed a 'fair working knowledge' of the group and offered to help any member wishing to take up their study. Huggins himself gives a list of some uncommon species in the district, such as *Sympetrum rubrum* (=*sanguineum?*), *S. flaveolum, A. imperator* and *A. mixta*. Unfortunately, Huggins' efforts appear to have borne little fruit, as subsequent issues of the *Bulletin* carried no further dragonfly reports. Nevertheless, Benfleet continued to be visited by London-based and west Essex entomologists, and it was on one such excursion of the Entomological Section of the LNHS, in search of the elusive *Lestes dryas* that, in July 1946, *Coenagrion scitulum*, a species new to Britain, was discovered. Pinniger and Longfield were both responsible for the initial discovery and subsequent identification, whilst Hammond was called upon to do the first illustrations.

By the late 1940s, sufficient field work across Essex had been conducted for an assessment of dragonfly distribution to be made, as distinct from presenting lists for specific localities. The first of these more ambitious undertakings was Cynthia Longfield's very substantial article on 'The Dragonflies (Odonata) of the London Area' (C. Longfield 1949) in the *London Naturalist*. As a professional biologist, based at the London Natural History Museum, Longfield had a long and distinguished career, and an influence on the study of dragonflies of international importance. But from the point of view of the present study, it was her active involvement in local and amateur natural history circles, and the help and encouragement she gave to many of our most committed field naturalists which should be recognised. Along with Syms, Hammond, Pinniger and others, she was a fellow of the Royal Entomological Society, and could recall the excitement generated by the exhibition of two specimens of *C. scitulum* at a monthly meeting.

As well as her activity in the more professionally-oriented national scientific societies, Longfield found time and energy to play an active role in the LNHS over many years before returning to live in Ireland in 1957. She joined the Society in 1926, was soon active in the newly-formed Entomological section, and became the first woman president of the Society in 1932. She continued for a second year of office, and her presidential addresses were characteristically wide-ranging, on the history of natural history, and on her overseas natural history expeditions. She was particularly active in the Entomological Section, chairing it for eight years prior to 1951. She subsequently took the chair of the Nature Conservation Committee of the Society. She was keen to encourage other naturalists, including the young E.B. Pinniger and she recollected how 'he often brought his finds to me at the Museum' (pers. comm.).

Longfield's paper on the dragonflies of the London area was prepared as one of a series of surveys initiated by the Entomological Section of the LNHS, and she acknowledged the co-operation of L. Parmenter, R.M. Payne, E.B. Pinniger and C.O. Hammond among others. Her remarks on Essex were mainly confined to that part of the west of the county which lies within the LNHS area (20 miles radius of St Paul's Cathedral). However, since this area includes some of the most important and (by previous naturalists) best-studied sites in the county, the value of the report is considerable.

Longfield listed as still present in the LNHS section of Essex some twenty-five species (including the two rare migrant *Sympetrum* species). She also listed a further seven species as having been recorded in this part of the county previously, but subsequently lost. These seven included *C. pulchellum*, which, she commented, had not been known in the county since the end of the 19th century, until its discovery at Foulness in 1943. In view of the inclusion of this species in both Syms' and Pinniger's lists for Epping Forest, with which Longfield was undoubtedly familiar, this comment is surprising. Given the variability of *C. pulchellum* and the great difficulty of separating some of its forms from *C. puella* it may be that some doubt persisted over the correct determination of some records of the former species.

With the somewhat doubtful exceptions of *C. pulchellum* and *O. coerulescens* (by then only recorded once from Essex in any case) it appears from Longfield's list that no species had been lost from that part of Essex covered by the survey in the previous half-century. One or two, such as *C. virgo*, were said to be in decline, whilst two more, *O. cancellatum* and *A. mixta*, had established themselves as new breeding species in the county. Yet another had been recorded for the first time (Pinniger's capture of a single specimen of *C. boltonii* in 1930 in Epping Forest). Essex sites mentioned by Longfield include the Lea Navigation Canal, the Rivers Roding and Stort, Epping Forest,

Coopersale Common, Chingford, Ongar Park, Hainault Forest, and reservoirs at Wanstead and Walthamstow, within the LNHS area, and Foulness beyond it.

Another first rate amateur field naturalist who did significant field work on the Essex dragonflies during the 1940s was the late Bernard T. Ward. R.M. Payne regards him as 'the best all-round naturalist Essex produced this century' (pers. comm.). In an interview published in 1971 (Ward 1971) Bernard Ward described the breadth of his interests: 'Whilst botany has been my main interest, I have delved into many disciplines in the natural history field, particularly ornithology, bryophytes, odonata, some groups of diptera, mycetozoa, lepidoptera and phytophagous hymenoptera. In brief, my interests have included anything concerned with natural history and local history too'. Ward knew the Essex countryside intimately, depite (or perhaps because of?) never having owned a car, and Pinniger (who knew him for some fifty or sixty years) recalled many excursions together, in later years accompanied by their wives.

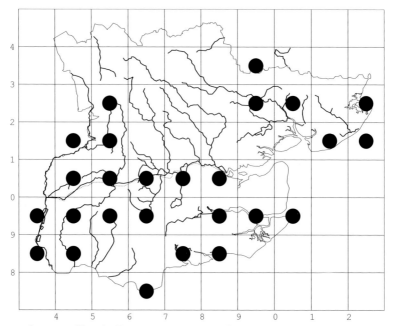

Dragonfly recording in Essex – coverage prior to 1950

Like Pinniger, Ward (who also lived at Chingford) was encouraged in his early natural history interests by his parents (and by a local gardener). His interest was also stimulated when he met, while walking in Epping Forest, a member of the local natural history society, but it was not for another eleven years that he was able to afford the membership fee for the Essex Field Club. In fact he seems to have joined the LNHS first (he is referred to in the *London Naturalist* for 1928), and was editor of the *London Naturalist* from 1933 to 1937 (the period during which Pinniger's dragonfly reports appeared) and later chaired the Chingford branch.

Ward joined the Essex Field Club in the early 1930s and subsequently became Joint Hon. Secretary, Honorary Member, President (1964–6), and Curator of the Epping Forest Museum (1948–61). He published a series of botanical notes between 1957 and 1959 in the *Essex Naturalist*.

In later life, Ward recalled his indignation at seeing long rows of set specimens when he first joined an (un-named) natural history society. He was always a conservationist, and took a leading role in the formation of the Essex Naturalists' Trust in 1959 (he was Chairman of the Trust for 1960–61). He was a verderer of Epping Forest for many years and E.B. Pinniger comments: 'His work for and in Epping Forest will I hope be remembered by those who enjoy walking in the Forest' (pers. comm.).

Some of Bernard Ward's dragonfly specimens, dating from 1940, 1949 and 1950 were retained at the Epping Forest Museum (Queen Elizabeth's Hunting Lodge). They are mainly common species, and come from Hatfield Forest, Grays and Fyfield, as well as from Epping Forest sites. The *Essex Naturalist* vol. 28 (covering the period 1945 to 1950) contains a number of references to field excursions (to Margaretting, the Stour Valley, Stanford Rivers and Toot Hill) with dragonfly lists provided by Bernard Ward. Some of these records were subsequently incorporated in an article under the joint authorship of Pinniger, Syms and Ward in the same volume, and titled 'Dragonflies in 1949'. Though by no means a comprehensive county survey, the article was probably the nearest approximation to one produced by that time. A total of twenty-two species was recorded from various sites throughout the county by the three authors. If we discount the rare migrant *Sympetrum spp.*, and the single occurrence of *C. boltonii*, the list is almost identical to that given by Longfield for the LNHS corner of Essex. The presence of *Coenagrion scitulum* in south-east Essex (reported by several observers) is a significant difference between the lists, as is the fact that none of the authors appear to have seen

C. *virgo* in 1949, despite visits to sites where it might at that time have been expected to turn up. This seems to confirm (as has subsequent experience) Longfield's judgment that this species was in decline in Essex.

The localities listed by Pinniger, Syms and Ward make an impressive list for one season: Connaught Water, Chingford Plain, Monk Wood, Cuckoo Pits, Strawberry Hill Pond, Baldwins Hill Pond and Fairmead Pond in Epping Forest, Hainault Forest, Hatfield Forest, the Stort near Little Parndon, Curtis Mill Green, Ongar, Fyfield, Stanford Rivers and Shelley in the Roding Valley, Warley and Childerditch Commons, near Brentwood, Chingford, Coopersale Common, Wanstead Park, Margaretting and Benfleet are all named. Though significantly biased towards the west of the county, this is a far wider geographical spread than is represented in any earlier list.

1950–1977: waiting for Hammond

Interest in dragonflies seems to have faded somewhat from the early 1950s, not to be revived until the late 1970s, with the publication of C.O. Hammond's beautifully illustrated *The Dragonflies of Great Britain and Ireland* (Hammond 1977). Essex was of national importance by virtue of the presence of C. *scitulum* and as a stronghold for *Lestes dryas*. For this reason, south-east Essex, and the Thames estuary continued to receive regular attention from Odonatists and others. Cynthia Longfield recalls (pers. comm.) visiting the C. *scitulum* site every year until she left Britain in 1957 (though the species itself was not seen after the floods of 1953). She also said that L. Parmenter and R.M. Payne continued the search without success. *Lestes dryas* also continued to be the subject of some interest, the Benfleet site being regularly visited, as well as new sites discovered (e.g. South Woodham Ferrers, 1950 and Flatford Mill, also 1950 (R. Merritt, pers. comm.)). Hammond, in particular, continued to visit Benfleet into the 1970s and his diaries record an impressive total of eleven species. *Lestes dryas* was last seen there by him in 1971. In the wake of fears that *Lestes dryas* might have become extinct in Britain, there were several visits to known and possible L. *dryas* sites in coastal and estuarine Essex in the mid-1970s. At that time, the decline of another local species, S. *sanguineum* was giving cause for concern.

Sporadic recording of the Essex Forest Odonata was also continued by a variety of other naturalists, including Hammond himself, B.T. Ward, J. Owen-Mountford and several others until the late 1970s when systematic work began again with the valuable studies of the late E.P. Ryan from 1979. In

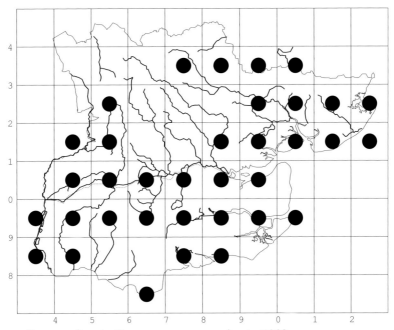

Dragonfly recording in Essex – coverage prior to 1980

the north-east of the county, the re-establishment of the Colchester Natural History Society, with its associated local publications, provided a context for some work on dragonflies to be recorded. The great Suffolk naturalist and photographer, S. Beaufoy is reported as having given a lecture on Dragon-flies to the Society in February 1960, and the same annual report (for 1960) carried a record by J.A. Richardson of a single male of the local ruddy darter (*S. sanguineum*) at Rowhedge sandpits, seen in October 1957.

Also in the Colchester area, the newly established gravel-pit reserve of the Essex Naturalists' Trust at Fingringhoe Wick was the subject of some sur-vey work by members of the CNHS. The Report for 1962 and 1963 contains another piece by Richardson, in which he gives *A. cyanea* and *L. depressa* as species occurring at the Wick (Richardson 1963).

In all, at least some dragonfly records had been obtained from approximate-ly thirty-six out of a possible fifty-seven 10km squares in Essex by the end of 1979.

The 1980s

Stimulated by the ready availability of a reliable and reader-friendly identification guide in the shape of Hammond's *Dragonflies of Great Britain and Ireland*, numerous observers began to study the group in a systematic way from the end of the 1970s. J. Shanahan, E.P. Ryan, J. Dobson, T. Benton, D. Smith and others were recording dragonflies in various parts of the county and soon made contact with each other. From the beginning of the 1980s this group expanded and a systematic survey of the county's dragonflies was co-ordinated, incorporating the records of over seventy observers. This resulted in this work's predecessor, *The Dragonflies of Essex*, published by the Essex Field Club in 1988.

Species formerly recorded in Essex, but not recorded during the present survey

A. *Species reported in Essex prior to 1900, but not recorded since:*

The reports listed here are drawn from published sources, but it should be remembered that settled terminology for dragonfly species and reliable information about which species occurred in Britain were not available until the 20th century. As we note below, several of these records are of doubtful validity, but one or two have become more plausible in the light of recent discoveries.

Ceriagrion tenellum (de Villers, 1789) – Small Red Damselfly

In Britain, this species occurs at the north-western limit of its European range and is most frequently found in sphagnum bogs and wet heathland in southern England and Wales. Although this habitat is increasingly threatened, *C. tenellum* may be locally abundant where suitable habitat exists. The species was said to occur commonly in Cambridgeshire (Morley 1929), but became extinct in that county before the end of the 19th century. It was first recorded in Suffolk in 1945, and a second site was located in 1949 (Mendel 1992). However, both sites were lost and the species was last recorded from Suffolk in 1950. Edward Doubleday (1835) described it (as *Agrion rubellum*) as occurring 'in profusion' in the vicinity of Ongar Park Woods, near Epping. Subsequently, *C. tenellum* (as *Pyrrhosoma tenellum*) was described as 'very common formerly' at Coopersale Common on Henry Doubleday's (1871) list. The use of the word 'formerly' may indicate that the species was already extinct and although later lists (Lucas 1902; Harwood 1903; Longfield 1949) all quoted Doubleday, they offered no further records. Given the lack of suitable habitat in the county, it is unlikely that this species survived in Essex after 1900 – and may have already been extinct by the time of Doubleday's 1871 list.

Coenagrion mercuriale (Charpentier, 1840) – Southern Damselfly

This species, which is protected under the Wildlife and Countryside Act, is restricted to around sixty sites in southern England and Wales with strongholds in the damp heaths of the New Forest and Dorset, as well as parts of south-west Wales. It is at the extreme north-western edge of its European distribution in Britain, and has specialised habitat requirements: typically small streams in heaths and bogs. The population in the New Forest was known in the 1890s and, at that time, the only other locality was considered to be Epping Forest, based on a specimen claimed to have been taken by Mr W.H. Nunney (Harcourt Bath 1892). No other reference is known for this record and it is not repeated in later lists. The specific habitat requirements for this species and recent work on its ecology (for example Evans 1989; Kerry 2001) make it seem unlikely that *C. mercuriale* has ever occurred naturally in Essex.

Ischnura pumilio (Charpentier, 1825) – Scarce Blue-tailed Damselfly

This uncommon species is found at scattered sites in southern Britain and is often associated with peaty runnels and bog seepages. It will also colonise mineral extraction sites where shallow pools in the early stages of plant succession occur. It is easily confused with the more widespread *I. elegans* and may be overlooked where this species is abundant. Henry Doubleday (1871) described it as 'rare', being occasionally found among old gravel pits around Epping – a record that was repeated without comment by both Lucas (1900b) and Harwood (1903).

However, Longfield (1949) considered that the habitat was 'a very peculiar one' for *I. pumilio*, and that the record was 'open to doubt'. Single 19th century records also exist for Norfolk (O'Farrell 1950) and Cambridgeshire but there are no confirmed records for Suffolk (Mendel 1992). *I. pumilio* was recognised in a chalk pit in Bedfordshire in 1975 by Nancy Dawson, but initially dismissed. In 1987 the species was re-discovered at two Bedfordshire sites, and at a similar locality in Buckinghamshire the following year (Cham 2004). In view of this the veracity of Doubleday's records cannot be ruled out.

Lestes virens (Charpentier, 1825) – Small Emerald Damselfly

Widely distributed in central and southern Europe, this species is a potential colonist from The Netherlands. Although not on the British list, it was reported by Henry Doubleday (1871) as 'rare among gravel pits', a record that was greeted sceptically by R. McLachlan, who subsequently wrote (McLachlan 1884) that the species was doubtfully British. Aware that he was casting doubt upon 'the late Henry Doubleday's discrimination' he pointed out 'the great difficulty that often attends the determination of the species of *Lestes*'. W.H. Harwood (1903) repeated McLachlan's opinion that the species was 'doubtfully British', pointing out that its only claim to appear in his list was that 'Mr Doubleday believed several specimens had been taken by himself at Epping'.

Gomphus vulgatissimus (Linnaeus, 1758) – Club-tailed Dragonfly

A distinctive dragonfly, *G. vulgatissimus* is locally distributed along a few slow-moving rivers (Severn, Thames, Wye, Arun) and their tributaries. In Essex, it was first documented in Edward Doubleday's (1835) list with the only subsequent record given by Henry Doubleday (1871), who reported it as 'very common at High Beech, and occasionally seen at other places near Epping' – a record repeated by Lucas (1900b), Harwood (1903), Pinniger (1933) and Longfield (1949). Although the late E.B. Pinniger (pers. comm.) recalled a conversation with W.J. Lucas many years earlier in which both expressed scepticism about Doubleday's records, his later experience of the species on the Thames led him to give more plausibility to the Doubleday records. Certainly Kemp and Vick's (1983) observations of the species on the Thames and Severn indicate a strong tendency of newly emerged individuals to disperse away from water and to make use of nearby woodland and scrub up to 10km from their site of emergence. Doubleday's sightings of the species at High Beech could indicate a past breeding colony on the Lea Navigation, a mere three kilometres or so distant.

Oxygastra curtisii (Dale, 1834) – Orange-spotted Emerald

Only recorded from the Moors River in Dorset in 1820–1963 and the River Tamar in 1946, *O. curtisii* is now considered extinct in Britain despite regular searches for it during the last sixty years. Henry Doubleday (1871) believed he saw a specimen of it, which he failed to catch. R.M. McLachlan (1884) subsequently commented that Doubleday might have seen *Somatochlora flaveomaculata* (a European species still to be recorded in Britain). Doubleday's identification is almost certainly mistaken, possibly a confusion with *G. vulgatissimus*, or even *C. aenea*, although he would have been very familiar with the latter species.

Leucorrhinia dubia (Vander Linden, 1825) – White-faced Darter

This attractive dragonfly of bog pools has been lost from half its English sites in the last fifty years (Smallshire & Swash 2004), mainly due to habitat destruction. According to Longfield (1949), J.F. Stephens recorded it at Epping some time before 1845. Her source for this comment is the remark by W.F. Evans (1845) that 'Mr Stephens states that it has been found at Epping' although Stephens (1835–7) does not give Epping as a locality for the species. A specimen was taken by a Mr Marsh in Epping Forest in 1841 (E. Doubleday 1841 and H. Doubleday 1841). It subsequently appeared on Henry Doubleday's (1871) list as 'always rare' at the old gravel pits on Coopersale Common. Longfield also noted that one of Doubleday's (presumably Henry) specimens caught on Coopersale Common in 1843 is in the Dale collection at Oxford. It seems certain that the species did exist in the vicinity of Epping in the mid-19th century and equally certain that it no longer exists in Essex. There are no subsequent records.

B. Species reported since 1900, but not re-found during the present survey

Coenagrion scitulum (Rambur, 1842) – Dainty Damselfly

E.B. Pinniger and C. Longfield, who were searching dykes near Benfleet in search of *Lestes dryas*, first discovered this species, now considered to be extinct in Britain, on 21st July 1946. Pinniger caught a male *Coenagrion* that neither of them could identify and subsequently, two females were found around three-quarters of a mile away. Both ditches were stagnant, in some parts brackish, and overgrown with spiked water-milfoil and sea club-rush. Longfield took the specimens back to the British Museum where they were photographed by W.H.T. Tams and painted by C. Hammond. From material and literature present at the Museum, Longfield was able to eliminate the closely related *C. hastulatum* and *C. coerulescens* becoming convinced that the specimens were *C. scitulum*, a species new to Britain. Contacts in Belgium, France and Italy were asked for specimens of *C. scitulum* to be sent for confirmation, although a complication at this stage was that the first male to be captured had an atypical marking on the second segment.

The first specimens were shown at meetings of the Royal Entomological Society, the South London Entomological and Natural History Society and the

London Natural History Society and donated to the national collection at the British Museum. Longfield reported the discovery in the *Entomologist* for 1947 and Pinniger published accounts in the *London Naturalist* for 1947 (Pinniger 1948) and the *Essex Naturalist* in 1948. This account was also able to include details of a male taken by Hammond on 21st June 1947 which also had atypical markings on the second abdominal segment. Longfield had already expressed the view that the specimens had bred locally (the *Entomologist*, vol 82, 1949, p.109 mistakenly included 1948 records of the species amongst its list of immigrant insects for that year), and Hammond's rediscovery in 1947 seemed to confirm this view (Hammond 1947).

The *Entomologist* for 1948 (Hammond 1948) carried a report from Hammond of another pair of *C. scitulum* seen on 22nd May of that year at the known locality. Since no others were seen, however, Hammond was unconvinced that this was the 'headquarters' of the species. This comment casts doubt on Hammond's later recollection (Hammond 1983, p.70) that he found the headquarters of the species in 1947. A.E. Gardner (1950) collected yet another male and female of *C. scitulum* on 19th June 1949, in the hope of studying the life-history if the species. Unfortunately, he was unable to get the female to oviposit. Presumably also in 1949, Hammond was able to donate two females to F.C. Fraser, who observed egg-laying four days later (14th July on water plantain). Nymphs were first observed on 28th August, and the final instar was entered during the first week of March the following year. Fraser subsequently published his account with detailed drawings of the larval stages (Fraser 1950). By 1949, the site for *C. scitulum* was known to a number of entomologists, and Pinniger, Syms and Ward (1950) reported sightings by several observers between May and July. Specimens collected on 18th June 1950 were exhibited by B.P. Moore at the SLENHS, and also by 1950 the main breeding site of the species must have been discovered, since Gardner (1954) reported having found larvae there early in 1950 and examined over a dozen adults later in the year.

In 1951 Gardner visited the main breeding site, a pond near Hadleigh (some two miles east of the original discovery), with common water-crowfoot and was able to examine over 250 individuals. He noted the great variability of the markings in the male (especially abdominal segments 2 and 9) and female (especially abdominal segment 9) and described these findings in the *Entomologist's Gazette* in 1952. Gardner revisited the site on 22nd June 1952, but found only seventeen individuals, including, however, two typical males (Gardner 1953).

Then came the disastrous east coast floods of 1953 when Canvey Island and the surrounding area were hit by three tidal waves which breached the sea walls. On 26th July, Gardner visited the site, but found the marginal vegetation badly grazed by cattle, the common water-crowfoot gone, and the pond, now very saline, covered by a filamentous alga. A few adults and one larva of *Ischnura elegans* were the sole surviving Odonata (Gardner 1954). Longfield also mentioned the effects of the flooding on the *C. scitulum* colony in her dragonfly report for 1953 (Longfield 1954). Noting that Gardner had fortunately been able to study its life-history, she concluded: 'It is too early to say that no remnant has escaped the floods, but none were seen in the summer of 1953 and there does not seem to be any other suitable area in the district, in which it could have survived. All of its known habitats were under the floods'.

Despite many subsequent searches, no further specimens of *C. scitulum* were recorded from the area. Gardner (1954) speculated that the original colonists were either ship- or wind-borne migrants. Since the area was already familiar to entomologists, it seems unlikely that it had been overlooked.

Coenagrion pulchellum (Vander Linden, 1825) – Variable Damselfly

This species is often difficult to distinguish from its more widespread and common relative, the azure damselfly (*C. puella*). The males of *pulchellum* have a 'stalk' to the black marking in abdominal segment 2 (giving a 'y' rather than 'u' shape, as in *puella*). The antehumeral stripes are also incomplete in *pulchellum*, giving them a '!' shape, and the black marking in segment 9 is more extensive than in *puella*. The females have two colour forms and are difficult to distinguish from those of *puella*. The most reliable feature is the shape of the pronotum ('collar'), which has a more deeply indented hind margin in *pulchellum*.

This species has an oddly discontinuous distribution in England and Wales, but is more common and widespread in Ireland.

Because of the possibility of confusion between this species and *C. puella*, the past history of this species in Essex is hard to interpret. Doubleday (1871) said it was 'not uncommon about the ponds by the side of the new road through the Forest', but it does not appear on the later lists compiled by the Campion brothers or by Hammond. However, Syms (1929) included it, as did Pinniger (1933): 'occurs rarely throughout the Forest'. Although he did not include it in his London Naturalist reports for 1934–8, Pinniger said it was present in the Forest until the 1940s (pers. corr.). A report of the species at Wanstead Park by F.A. Walker (1897) was not cited by either Lucas (1900b) or Harwood (1903), possibly indicating some scepticism about the record. Longfield believed it had not been seen in the county since Doubleday's list, until it was reported from Foulness in 1943. E.T. Levy subsequently reported it as occurring on flooded pits at Hawkwell in 1976. M. Chinery observed it in 1970 on the Suffolk side of the Stour and in nearby meadows. There is a further report from the River Lea on 10th August 1983 (B. Eversham, pers. corr.), when small numbers of adults and several exuviae were seen.

Despite careful searches we have been unable to confirm the continued presence of this species in the county, but there is a significant possibility that a breeding population may be present and overlooked.

Aeshna juncea (Linnaeus, 1758) – Common Hawker

Aeshna juncea is a species of wet heathland and moorland and may be found on some of the Surrey heaths although it is more common in western and northern Britain. This species is easily confused with *A. mixta* and *A. cyanea* and the misleading (for Essex) vernacular name has, on occasions, led to *A. mixta* being mistaken for *A. juncea*, and consequently, a number of reports of the latter have to be treated with some scepticism.

The common hawker appeared on none of the earlier lists for Essex, and Longfield (1949) did not list it as having been recorded in the county. She referred to its presence in Surrey, but within the LNHS boundary, she believed that it only occurred as a migrant. Though there have been a few reported sightings (G. Foott 1967, Fingringhoe Wick; 1953–8, Distillery Pond, Colchester (Colchester & Essex Museum)) from the east of the county, these are of doubtful validity. The late E.P. Ryan, an experienced observer, reported one hawking over Wake Valley pond in Epping Forest on 14th August 1983 (pers. comm.), an area where other scarce heathland species have occurred (see *S. danae* and *O. coerulescens*). This may be the only credible record for this species in Essex.

Cordulegaster boltonii (Donovan, 1807) – Golden-ringed Dragonfly

This large and distinctive dragonfly is commonly found in heathland and moorland habitats in northern and western Britain, including both Wales and Scotland. It is absent from East Anglia, the nearest populations being in Surrey, Sussex and West Kent, and it is presumed that it is wanderers from these areas that are occasionally recorded in Essex. The first of these known to us was an individual caught by E.B. Pinniger near Loughton in July 1930 (Pinniger 1933) with a later record of a male picked up alive in Romford, 1983 by P. Faithfull who retained the specimen. There have been several other reports of sightings of singletons (P.A. Smith, Epping Forest, 1970, B.R.C. and C. Plant, Claverhambury, 1980, pers. comm.) but despite subsequent visits to these sites, no further sightings were achieved in what appears to be unsuitable habitat. It is most unlikely that this species has ever been a resident, breeding species in the county.

Sympetrum vulgatum (Linnaeus, 1758) – Vagrant Darter

The vagrant darter is a rare migrant to Britain, most frequently recorded during influxes of other darters, notably *S. flaveolum* and *S. fonscolombii*. It is very similar to *S. striolatum*, but the black 'moustache' band on the face is continued downwards at the sides, along the edges of the eyes (not continued downwards in *striolatum*). The females have a prominent vulvar scale, projecting almost at right-angles below abdominal segment 8 (visible, but projecting at a more acute angle in female *striolatum*). As in *striolatum*, there is a thin yellowish line along the legs (black legs in *sanguineum*). Prior to the significant 'invasion year' of 1995 this species had been reliably recorded on less than ten occasions in Britain (Merritt *et al.*, 1997). These were mainly from the London area, with just one each from Devon and Yorkshire. Until that time there was one record only from Essex: a single male recorded by F.W. Campion and K.J. Morton in Epping Forest on 4th September 1906 (Campion & Campion 1906).

However, during the summer of 1995, a significant immigration of *Sympetrum* spp. occurred, amongst which at least fifteen *S. vulgatum* were recorded from three sites in Norfolk, with others in Suffolk (two sites), Kent and the London area. It seems likely that others may have been overlooked due to this species' close similarity to *S. striolatum* from which it can be separated only by close examination. One was caught and checked by N. Cuming at Colne Point on 9/08/95, and there is a report of another photographed by I. Rose at Martin's Farm, St. Osyth on 9/08/95 (pers. comm.). The species is widespread in central and north-eastern Europe, including the Netherlands. Careful scrutiny of darter dragonflies along the Essex coast, especially during periods of strong immigration by other *Sympetrum* species could well produce more records.

Sympetrum danae (Sulzer, 1776) – Black Darter

This was a resident species in Epping Forest at least until the late 1940s. There have been occasional sightings since then, but not during the period of the current survey.

It is a relatively small species of 'darter' dragonfly. The mature males are mainly black in colour, with paler, yellowish patches on the sides of the abdomen, and black legs. Females and immature males may be yellowish with small black markings towards the tip of the abdomen and a black triangular marking on the dorsal surface of the thorax.

The male is quite distinctive, as our only black 'darter' dragonfly. The black triangle on the thorax in immature and female specimens is a useful identifying feature.

S. danae is widespread and can be locally common where there is suitable habitat throughout northern and central Europe. Characteristic habitat is shallow acid pools in heaths and bogs or moorland, but, like several other *Sympetrum* species, it is liable to disperse considerable distances from its breeding sites, and is also an irregular migrant to Britain.

Occurrence in Essex

This species has been recorded periodically in Epping Forest between 1841 and the late 1940s. It seems likely that the species bred in the Forest for much of that period, though in fluctuating, often very low numbers. Discontinuities in reports may indicate lack of systematic recording activity, but it seems likely that the Epping population may have itself died out periodically, with recurrent recolonisations.

The first report of which we are aware is a note by Henry Doubleday to the *Entomologist* dated 16th September 1841. He lists *danae* (as 'scoticum') as one of three species (of *Sympetrum*) which suddenly appear in profusion at a spot where they have not been seen in previous years. In Doubleday's 1871 list, *danae* is reported to be 'common in certain years among the old gravel pits'. The Campion brothers did not see it until 1906, when it was seen at two separate sites in the Forest. It appeared on E.E. Syms' 1929 list, and according to E.B. Pinniger (1933) it was 'always present in the Forest, though numbers fluctuate considerably from year to year'. In fact, only one was seen in 1933

(Pinniger 1934). Hammond's notes include no reference to the species before 1926, and even then 'it never became common'. There are further reports in Hammond's notebooks for 1944 and 1945. It was mentioned in the report by Pinniger, Syms and Ward (1950) as present 'in some numbers' at one pond in the Forest. Longfield (1949) says it was known from the Forest 'before 1845 and has been found there ever since, but is never common'.

We know of no sightings of this species in Epping Forest or elsewhere in the county between 1950 and 1980. More systematic searches during the 1980s produced no new records until a single male was observed in the Forest on 16th August 1987 by A. McGeeney.

This was followed by a scattering of reports of single individuals, presumably migrants: one at West Wick on the Dengie Peninsula on 12th October 1991 seen by D. Wood, and another on Old Florrie's green lane (TL8625) near Coggeshall on 30th June 1994, seen by U. Broughton.

During the major *Sympetrum* 'invasion' year of 1995 several *danae* were reported as present among the other migrant species seen at Yarmouth cemetery on 5th August, and others were seen at various Suffolk and Kent sites (Pittman 1996, Silsby 1995). For that year we have only one Essex report: a male seen by A. McGeeney at Speakman's pool in Epping Forest in late August.

Significant numbers of migrant *danae* were reported along the east coast of Britain in August and early September 1999 (Parr 2000), and this may have been the source of one (or more?) males seen in Epping Forest during that period:

One seen on 28th August at 11.00am. at Long Running, Epping Forest (TQ433997): perching and hunting over shallow pond in warm bright sun. (P. Coghlan). J. Stevens reported one observed on 30th August at the same site 'holding territory' by an unfenced pond at the north end of the cattle enclosure. On 2nd September a male, together with what might have been a female, was seen at the same site by A. Spicer & T. Gray. Presumably it was the same male seen on each occasion.

Possible future visitors and colonists: species to look out for

The rapid colonisation of Essex and then other parts of the south-east by the small red-eyed damselfly could mark the beginning of significant changes in our dragonfly fauna. The rare migrant *Sympetrum* species are now reported more frequently, and we now have an Essex breeding record of one of them: *Sympetrum fonscolombii*. The yellow-winged darter (*Sympetrum flaveolum*) has established short-lived breeding populations elsewhere in Britain, and there is no obvious reason why this should not happen in Essex.

Similar considerations also apply to the lesser emperor (*Anax parthenope*). Elsewhere in Britain there have been recent records of the scarlet darter (*Crocothemis erythraea*) and the banded darter (*Sympetrum pedemontanum*). Both these species appear to be expanding their range in the European mainland, and future colonisation of Britain is a distinct possibility.

It is also possible that several of the species mentioned in **Appendix 1** may subsequently be discovered in Essex, or may re-colonise. A close watch should be maintained for *C. pulchellum*, particularly in the Lea Valley and along the River Stour. There is also an outside possibility of the heathland species *S. danae* and *O. coerulescens* establishing themselves in the Epping Forest area: the chances of this could be greatly enhanced by suitable extension and management of areas of damp heathland in the forest.

There are several species that are common on the European mainland that appear 'poised' to establish themselves in Britain. These include the common winter damselfly (*Sympecma fusca*), the dainty damselfly (*C. scitulum* – see **Appendix 1**), and perhaps the willow emerald (*Lestes viridis*), of which there is one breeding record from north Kent (Brooks in Brooks (ed.) 2004), and the goblet marked damselfly (*Erythromma lindenii*).

Perhaps the most likely future Essex colonist is the southern emerald (*Lestes barbarus*).

Lestes barbarus (Fabricius, 1798) – Southern Emerald

This species is widespread throughout southern and central Europe, more common in the south, but recently extending its range northwards. It is a rare immigrant to the east coast of Britain, first noticed at Winterton Dunes, Norfolk by G. Nobes on 30th July and 7th August 2002. There were further reports in 2003 from Winterton and also from Sandwich Bay, Kent. In 2004 up to 15 individuals were seen at Sandwich Bay between 15th July and the end of August, and ovipositing pairs were observed. Small numbers of males were also seen again at Winterton, but during a period of migration. It is possible that breeding was established briefly at Sandwich Bay, but if so inundation of the site with sea water in January 2005 may have put an end to the species there. There were no records in 2005.

There are strong populations in coastal dunes in the Netherlands, and it seems possible that the species may appear again on the east coast. Shallow ponds in marshes and dunes seem to be a favoured habitat, and suitable locations of the Essex coast – such as Colne Point – should be checked for this species.

Identification: There are wide yellow stripes on the thorax (unlike other *Lestes* species), and the pterostigma is white/yellowish in the outer part, brownish on the inner (uniform black/brown in the other *Lestes* species). The rear of the head is yellow in both sexes (green in other *Lestes* species). The male anal appendages are yellowish with dark tips (predominantly dark/black in the other species)

(Nobes 2003, Parr 2003, 2004, 2005, 2006).

Appendix 3
The discovery of the Small Red-eyed Damselfly in Essex

This article was originally published in the January 2000 issue of Atropos, *the journal for butterfly, moth and dragonfly enthusiasts, and is reproduced here with the kind permission of the editor and the authors.*

On 17th July 1999 whilst out looking for wildlife in an area of unspoilt countryside in Essex, a short detour to an unpromising looking pond was to lead to the discovery of a species of damselfly that had never previously been recorded in Britain.

On a lily pad some distance from the shore an unfamiliar damselfly could be seen sitting. Although it bore a superficial resemblance to **Red-eyed Damselfly** *Erythromma najas*, this insect seemed very small for that species and the blue markings on both head and abdomen were clearly quite different. Our thoughts immediately turned to **Small Red-eyed Damselfly** *E. viridulum*, a European species which had been featured in the April 1999 issue of *Atropos* (Wasscher 1999). After a while, a second individual appeared and the two insects (which were clearly the same species) were viewed both at rest and whilst flying out over open water, returning again and again to the same plants. That evening an examination of the available literature made us even more certain of what we had seen (*see Wasscher 1999 for a full discussion of the identification of this species. Eds*). Proof would rightly be needed, and in any case we had no way of knowing whether we had stumbled across a colony of a new resident British species, or whether we had been fortunate enough to observe a couple of primary migrants.

On 23rd July, armed with a camera, binoculars and butterfly nets, some of our questions were to be answered. A superb male was quickly located and photographed close to the western shore, and almost immediately a pair were seen in tandem. A third male was found settled on waterweed at the eastern edge of the pond and this individual was promptly netted. Under close examination, the damselfly was positively identified as Small Red-eyed Damselfly. After a selection of photographs had been taken it was returned to its habitat and safely released.

The fine weather in July was followed by a very wet August, and for the first three weeks good days for insect observation were depressingly few and far between. However, the colony was kept under observation and on 11th August at least four egg-laying pairs and three single males were found to be present. Best of all, a new colony was discovered over a quarter of a mile from the original, and this one certainly proved to be much stronger. Whilst ovipositing was repeatedly witnessed at both sites, numbers at the first faded quickly and the total of 11 on 11th August was never exceeded. On 15th August, several *E. viridulum* were located at yet another location, and during this fine spell in late August and early September this one was certainly the most impressive. Very large numbers were observed over the water and many were to be seen hunting amongst waterside bushes and trees, the majority of which were males. Indeed, the numbers here were so great that it was quickly established that Small Red-eyed Damselfly was the dominant species.

After an all too brief spell of dry and warm weather, heavy rain returned by mid–September and *E. viridulum*'s season was prematurely brought to a conclusion. With the last individuals observed on 21st September (two males), it is to be hoped that enough eggs have been laid for an apparently successful colonization to continue. An unsuccessful attempt was made to find exuviae, but we were probably much too late in the season.

It is worth noting that a large number of other possible breeding sites were checked over a ten to fifteen mile radius but no other individuals or colonies were located. Until such time as a foothold in this country appears more secure, the authors have decided to withhold information that could identify the existing localities. It is hoped that readers will understand and respect our decision.

Reference

Wasscher, M. (1999) 'Identification of Small Red-eyed Damselfly *Erythromma viridulum* (Charp.)' *Atropos* **7**: 7–9.

Lesser Emperor Dragonfly

Location: Bedfords Park (TQ515920), Essex.

Date: 8th August 2003.

Observers: Colin Jupp (finder) and Andrew Middleton.

Weather: hot, still and sunny day.

Purpose of visit: visit Bedfords Park with Colin, especially to view small red-eyed damselflies, and try for some images with telescope and digital camera.

Colin and I were on the south shore of the square fishing lake at around 12:15hrs where we had been watching the dragonflies for a while. I had just begun preparing my optical equipment to try for some images of small red-eyed damselflies on the floating vegetation, whilst Colin was still looking with his binoculars at the odonata flying over the water.

Then Colin said something like: 'Hey what's this', and so I knew it would be wise to raise my binoculars ASAP. There, perhaps 10m in front of us, and 1–2 metres above the water, was a strange 'hawker' type dragonfly, medium plus in size.

What was immediately visible was a bright blue band around the base of the abdomen adjacent to the thorax and spreading onto the constriction or waist of the abdomen near the thorax. As the flight direction was now rather away from me, although somewhat side-on, and below head height, I could see that the abdomen was coloured a dull and rather dark and dingy olive-brownish and along the top of the abdomen ran a rather darker line contrasting slightly with the sides. The abdomen appeared to be held rather straight, unlike the usual flight shape of patrolling male Emperors *Anax imperator*, of which there were several present.

My next check was on the thorax sides, which again were rather dark and dingy, although not quite as dark as the abdomen sides, and certainly not patterned, or greenish or yellowish. My next check would have been the eyes, but before this was accomplished, the insect disappeared from the lake.

This sighting lasted, I'm not sure, but perhaps 20–30 seconds. During this short time, I believe we both remarked that it was a Lesser Emperor. We would have liked further views to confirm further features and to enjoy the insect more, but were both completely happy with the identification, having such a distinctive insect in full view at c.10m, through binoculars with the sun behind us.

Some time after, I believe I saw the insect briefly around hawthorns over meadows just to the east of the lake, then around 12:45hrs, Colin had definite views of the Lesser Emperor around the bushes here, being chased, but I didn't see it this time.

We watched the area through the afternoon, but had no more luck with the Lesser Emperor, and I believe the lake was watched through the evening as well with negative results. The presence of several male Emperors *Anax imperator* patrolling the lake and its marginal habitats probably proved too much for the Lesser Emperor. The situation wasn't helped by the presence throughout of a kestrel, which would mount regular swoops from an adjacent bush over the lake, and would invariably take a larger dragonfly.

My field experience at this time related to the resident 'hawker' species found in Essex, plus golden-ringed and common hawker, and one southern migrant hawker (France). I was aware of the main identification features of lesser and vagrant emperor prior to the sighting, and had also seen several images of these species.

Scientific names of plants mentioned in the text

Amphibious bistort	*Polygonum amphibium*
Arrowhead	*Sagittaria sagittifolia*
Branched bur-reed	*Sparganium erectum*
Bulrush	*Schoenoplectus lacustris*
Burdock	*Arctium lappa*
Common reed	*Phragmites australis*
Common spike-rush	*Eleocharis palustris*
Canadian water-weed	*Elodea canadensis*
Hornwort	*Ceratophyllum demersum*
Lesser reedmace	*Typha angustifolia*
Pondweed	*Potamogeton* species
Pondweed, broad-leaved	*Potamogeton natans*
Pondweed, fennel	*Potamogeton pectinatus*
Purple loosestrife	*Lythrum salicaria*
Reedmace	*Typha latifolia*

Reed sweet-grass   *Glyceria maxima*

Sea club-rush   *Bolboschoenus* (formerly *Scirpus*) *maritimus*

Spiked water-milfoil   *Myriophyllum spicatum*

Water-cress   *Rorippa nasturtium-aquaticum*

Water crowfoot   *Ranunculus aquatilis*

Water horsetail   *Equisetum fluviatile*

Water lily   *Nuphar* and *Nymphaea* species

Water mint   *Mentha aquatica*

Water moss   *Fontinalis antipyretica*

Water plantain   *Alisma plantago-aquatica*

Water speedwell   *Veronica anagallis-aquatica*

Yellow iris   *Iris pseudacorus*

Bibliography

Aguilar, J.d', Dommanget, J-L. & Prechac, R. 1986 *A Field Guide to the Dragonflies of Britain, Europe and North Africa*. London: Collins.

Bailey, B. & Bailey, J. 1985 Terrestrial feeding by a larva of *Calopteryx virgo* (L.) *Journal of the British Dragonfly Society* **1** (5): 72.

Bailey, M.P. 2000 Predation of Four-spotted Chaser *Libellula quadrimaculata* L. by Otter *Lutra lutra* L. *Journal of the British Dragonfly Society* **16** (2): 64.

Bath, W. Harcourt 1890 *Illustrated Handbook of British Dragonflies*. Birmingham: Naturalist's Publishing Co.

Bath, W. Harcourt 1892 Some famous collecting grounds for dragon-flies. *Science Gossip* **28** (Pt.1): 17-18: (Pt.2): 65–6.

Bath, W. Harcourt 1893 Observations on British Odonata. *Entomologist* **26**: 108.

Benton, E. 1988 *The Dragonflies of Essex*. Stratford: Essex Field Club.

Benton, T. 1997/8 Dragonfly (Odonata) report for 1997. *The Essex Naturalist* **15 (New Series)**: 36–7.

Benton, T. & Dobson, J. 2002 Essex dragonflies (Odonata) an update. *The Essex Naturalist* **19 (New Series)**: 86–8.

Benton, T. & Dobson, J. 2003 Essex dragonfly report for 2002. *The Essex Naturalist* **20 (New Series)**: 60–1.

Benton, T. & Payne, R.G. 1983 On the rediscovery of *Lestes dryas* Kirby in Britain. *Journal of the British Dragonfly Society* **1** (2): 28–30.

Beynon, T.G. 1998 Behaviour of immigrant *Sympetrum flaveolum* (L.) at breeding sites in 1995 and subsequent proof of breeding in 1996. *Journal of the British Dragonfly Society* **14** (1): 6–11.

Brook, G. 2003 Identification of the exuvia of the Small Red-eyed Damselfly *Erythromma viridulum* (Charpentier). *Journal of the British Dragonfly Society* **19** (1 & 2)**: 40–3.

Brook, J. & Brook, G. 2003 The Willow Emerald Damselfly *Chalcolestes viridis* (Vander Linden) in Kent: a case of mistaken identity. *Journal of the British Dragonfly Society* **19** (1 & 2): 51–54.

Brooks, S.J. 1989 The dragonflies (Odonata) of London: the current status. *The London Naturalist* **68**: 109–131.

Brooks, S.J. (ed.) 2004 *Field Guide to the Dragonflies of Great Britain and Ireland.* Hook, Hampshire: British Wildlife.

Brooks, S.J., McGeeney, A. & Cham, S.A. 1997 Time-sharing in the male Downy Emerald, *Cordulia aenea* (L.) (Corduliidae). *Journal of the British Dragonfly Society* **13** (2): 52–7.

Brownett, A. 1990 Predation of *Enallagma cyathigerum* (Charpentier) by the Grey Wagtail (*Motacilla cinerea* Tunstall). *Journal of the British Dragonfly Society* **6** (1): 1–2.

Brownett, A. 1994 Resource partitioning in the genus *Calopteryx*: an unsolved problem of odonatology. *Journal of the British Dragonfly Society* **10** (1): 6–11.

Brownett, A. 1998 Predation of adult *Anax imperator* Leach by the Hobby (*Falco subbuteo* L.) – how frequently does this occur? *Journal of the British Dragonfly Society* **14** (2): 45–52.

Buck, F.D. 1966 Obituary: E. E. Syms, 1881-1966. *Proceedings and Transactions of S.L.E.N.H.S.* 1966: 131–2.

Campion, F.W. & Campion, H. 1903 The Dragonflies of Epping Forest. *Entomologist* **36**: 49–50.

Campion, F.W. & Campion, H. 1904a The Dragonflies of Epping Forest in 1903. *Entomologist* **37**: 19–20.

Campion, F.W. & Campion, H. 1904b On a dark form of *Ischnura elegans* (female). *Entomologist* **37**: 252–4.

Campion, F.W. & Campion, H.1904c The dragonflies of Epping Forest in 1904 *Entomologist* **37**: 30–1.

Campion, F.W. & Campion, H. 1905a *Aeshna mixta* in Epping Forest. *Entomologist* **38**: 24.

Campion, F.W. & Campion, H. 1905b On the dark form of *Ischnura elegans* (female). *Entomologist* **38**: 298–9.

Campion, F.W. & Campion, H. 1906a The dragonflies of Epping Forest in 1905 *Entomologist* **39**: 36–7.

Campion, F.W. & Campion, H. 1906b *Orthetrum coerulescens* in Essex. *Entomologist* **39**: 160.

Campion, F.W. & Campion, H. 1906c The dragonflies of Epping Forest in 1906. *Entomologist* **39**: 277–83.

Campion, F.W. & Campion, H. 1907 The dragonflies of Epping Forest in 1907. *Entomologist* **40**: 274–7.

Campion, F.W. & Campion, H. 1909a The dragonflies of Epping Forest in 1908. *Entomologist* **42**: 7–10.

Campion, F.W. & Campion, H. 1909b On the trimorphism of *P. nymphula* (female). *Entomologist* **42**: 178–80.

Campion, F.W. & Campion, H. 1909c Notes on dragonfly parasites (larval water mites). *Entomologist* **42**: 242–6.

Campion, F.W. & Campion, H. 1909d The dragonflies of Epping Forest in 1909. *Entomologist* **42**: 293–6.

Campion, F.W. & Campion, H. 1910 On the variations of *Agrion puella*. Linn. (Odonata). *Entomologist* **43**: 1–5.

Campion, F.W. & Campion, H. 1913 Notes on the dragonfly season of 1912. *Entomologist* **46**: 77–9.

Cham, S.A. 1999 Roosting behaviour of some British Odonata with notes on the Scarce Chaser *Libellula fulva* Müller. *Journal of the British Dragonfly Society* **15** (2): 58–60.

Cham, S.A. 2002a The range extension of Small Red-eyed Damselfly *Erythromma viridulum* (Charp.) in the British Isles. *Atropos* **15**: 3–9.

Cham, S.A. 2002b Mate guarding behaviour during intense competition for females in the Common Blue Damselfly *Enallagma cyathigerum* (Charpentier). *Journal of the British Dragonfly Society* **18** (1 & 2): 46-48.

Cham, S.A. 2003 Factors influencing the distribution of the white-legged damselfly *Platycnemis pennipes* (Pallas) in Great Britain. *Journal of the British Dragonfly Society* **19** (1 & 2): 15–23.

Cham, S.A. 2003 Small Red-eyed Damselfly *Erythromma viridulum* (Charpentier) records in 2002. *Atropos* **19**: 19–24

Cham, S.A. 2004a *Dragonflies of Bedfordshire*. Bedford: Bedfordshire Natural History Society.

Cham, S.A. 2004b Observations on an inland population of the Small Red-eyed Damselfly *Erythromma viridulum* (Charpentier) with notes on the first discovery of larvae in Britain. *Journal of the British Dragonfly Society* **20** (1): 31–4

Cham, S.A. 2004c Oviposition behaviour of the two British species of Red-eyed Damselflies *Erythromma najas* (Hansemann) and *E. viridulum* (Charpentier). *Journal of the British Dragonfly Society* **20** (2): 37–41.

Charpentier, T. von 1825 *Horae Entomologicae*. Wratislava.

Charpentier, T. von 1840 *Libellulae Europeae*. Lipsiae.

Chelmick, D.G. 1979 *Provisional Atlas of the Insects of the British Isles Pt. 7 Odonata. Dragonflies*. Institute of Terrestrial Ecology.

Chelmick, D.G., Hammond, C.O., Moore, N.W., & Stubbs, A. 1980 *The Conservation of Dragonflies*. Nature Conservancy Council.

Corbet, P.S. 1957 The life history of the emperor dragonfly *Anax imperator* Leach (Odon., Aeshnidae). *Journal of Animal Ecology* **26**: 1–69.

Corbet, P.S. 1999 *Dragonflies: Behaviour and Ecology of Odonata*. Colchester: Harley.

Corbet, P.S., Longfield, C. & Moore, N.W. 1960 *Dragonflies* (New Naturalist 41). London: Collins.

Cross, I.C. 1987 A feeding strategy of a Pied Wagtail (*Motacilla alba yarelli* L.) on *Libellula depressa* L. *Journal of the British Dragonfly Society* **3** (2): 36–7.

Dewick, S. & Gerussi, R. 2000 Small Red-eyed Damselfly *Erythromma viridulum* (Charpentier) found breeding in Essex – the first British records. *Atropos* **9**: 3–4.

Djikstra, K.-D.B. & Lewington, R. (ed. and ill.) 2006 *Field Guide to the Dragonflies of Britain and Europe*. Gillingham: British Wildlife.

Doubleday, E. 1835 Remarks on the entomology of Epping and its vicinity. *Entomological Magazine* 1835-6 (3): 147–59.

Doubleday, E. 1841 Note on *Sympetrum rubicundum* (= *L. dubia*). *Entomologist* 1st. Series 1841–2 (1): 159

Doubleday, H. 1841 Note on genus *Sympetrum*. *Entomologist* 1st Series 1841–2 (1): 205.

Doubleday, H. 1867 Note on emergence of Libellulae. *Entomologist* 1866–7 (3): 36.

Doubleday, H. 1871 A list of Odonata (dragon-flies) occurring in the neighbourhood of Epping. *Entomologists Monthly Magazine* 1871–2 (8): 86–7

Drake, C.M. 1990 Records of larval *Lestes dryas* Kirby in Essex during 1987. *Journal of the British Dragonfly Society* **6** (2): 34–41.

Drake, C.M. 1991 The condition of *Lestes dryas* Kirby larval populations in some Essex grazing marshes in May 1990. *Journal of the British Dragonfly Society* **7** (1): 10–17.

Dumont, H.J. & Hinnekint, B.O.N. 1973 Mass migration in dragonflies, especially in *Libellula quadrimaculata* L.: a review, a new ecological approach and a new hypothesis. *Odonatologica* **2**: 1–20.

Dunn, R.H. 1985 Some observations of *Aeshna cyanea* (Müller) ovipositing in unusual substrates. *Journal of the British Dragonfly Society* **1**: 99–100.

Evans, F. 1989 *A Review of Management of Lowland Wet Heath in Dyfed, West Wales*. NCC. Unpublished.

Evans, W.F. 1845 *British Libellulinae; or, dragonflies*. Private publication (printed by J.C. Bridgewater).

Firmin, J. *et al.* 1975 *A Guide to the Butterflies and Larger Moths of Essex*. Essex Naturalist's Trust.

Firmin, J. 2003 *Nature in North East Essex 2003*. Colchester: C.N.H.S.

Fitch, A. 1879 The past year. *Entomologist*. **12**: 281–291.

Follett, P. 1996 *Dragonflies of Surrey*. Woking, Surrey: Surrey Wildlife Trust.

Forsyth, L. 2005 *Island of Wildlife: the story of Fingringhoe Wick – a gravel pit nature reserve*. Colchester: Essex Wildlife Trust.

Gardner, A.E. 1950 *Coenagrion scitulum* (Rambur) in Essex. *Entomologist* **83**: 118.

Gardner, A.E. 1952a The life history of *Lestes dryas* Kirby (Odonata) *Entomologist's Gazette* **3**: 4–26.

Gardner, A.E. 1952b A note on the colour variations and separation characters of *Coenagrion scitulum* (Rambur) (Odonata) *Entomologist's Gazette* **3**: 161–6.

Gardner, A. E. 1953 Note on *C. scitulum*. *Entomologist's Gazette* **4**: 66.

Gardner, A.E. 1954 Is *Coenagrion scitulum* (Rambur) (Odonata, Coenagriidae) extinct in Britain? *Entomologist* **87**: 3–4.

Gibson, V. 2003 Communication between the sexes at the end of copulation: a study of three species of Anisoptera. *Journal of the British Dragonfly Society* **19** (1 & 2): 44–6.

Gibson, V. 2006 A study of the copulatory behaviour of three pairs of the Migrant Hawker *Aeshna mixta* Latreille in the wheel position. *Journal of the British Dragonfly Society* **21** (2): 47–54.

Goodyear, K.G. 2000 A comparison of the environmental requirements of larvae of the Banded Demoiselle *Calopteryx splendens* (Harris) and the Beautiful Demoiselle *C. virgo* (L.). *Journal of the British Dragonfly Society* **16** (2): 33–51.

Gunton, T. 2000 *Wild Essex*. Wimbish: Lopinga.

Hagen, H. 1857 A synopsis of the British dragon-flies. *Entomologist's Annual* 1857: 18–30.

Hammond, C.O. 1946 *Sympetrum flaveolum* and *S. sanguineum* at Wood Green and Richmond Park. *Entomologist* **79**: 60.

Hammond, C.O. 1947 Note on *C. scitulum*. *Entomologist* **80**: 272.

Hammond, C.O. 1948 Note on *C. scitulum*. *Entomologist* **81**: 284.

Hammond, C.O. 1977 *The Dragonflies of Great Britain and Ireland*. London: Curwen. Revised R. Merritt (1983) Colchester: Harley Books.

Harwood, W.H. 1903 Insecta. *The Victoria County History of the Counties of England: Essex*. **1**: 91–8

Hayter-Hames, J. 1991 *Madam Dragonfly: The life and times of Cynthia Longfield*. Durham: Pentland.

Hofmann, T.A. & Mason, C. F. Competition, predation and microhabitat selection of Zygoptera larvae in a lowland river. *Odonatologica* **34**: (1): 27–36.

Holmes, J.D. 1984 Rapid larval development in *Brachytron pratense* (Müller). *Journal of the British Dragonfly Society* **1** (3): 38.

Huggins, H.C. 1939 Entomological notes. *Bulletin of South Essex Natural History Society*. No. 27. Jan.

Jones, S.P. 2000 First proof of successful breeding by the Lesser Emperor *Anax parthenope* (Sélys) in Britain. *Journal of the British Dragonfly Society* **16** (1): 20–23.

Kemp, R.G. & Vick, G.S. 1983 Notes and observations on *Gomphus vulgatissimus* (Linnaeus) on the River Severn and River Thames. *Journal of the British Dragonfly Society* **1** (2): 22–5.

Kerry, L. 2001 Habitat management for the Southern Damselfly *Coenagrion mercuriale* (Charpentier) on Aylesbeare Common, Devon. *Journal of the British Dragonfly Society* **17** (2): 45–8.

Ketelaar, R. 2002 The recent expansion of the Small Red-eyed Damselfly *Erythromma viridulum* (Charpentier) in The Netherlands. *Journal of the British Dragonfly Society* **18** (1 & 2): 1–8.

Kirby, W.F. 1890a A revision of the subfamily Libellulinae, with descriptions of new genera and species. *Transactions of the Zoological Society of London* **12**: 249–348.

Kirby, W.F. 1890b *A Synonymic Catalogue of Neuroptera Odonata or Dragonflies: with an Appendix of Fossil Species.* London: Gurney & Jackson.

Leach, W.E. 1815 Entomology. *Brewster's Edinburgh Encyclopaedia*. 9pt. **1**: 57–172.

Linden, P.L.Vander 1825 *Monographiae Libellelulinarum Europaenum*. Bruxelles.

Longfield, C. 1937 *The Dragonflies of the British Isles.* London & New York: Warne.

Longfield, C. 1946 Further evidence of the immigration of *Sympetrum sanguineum* (Müller) together with *S. flaveolum* (L.) and *S. striolatum* (Charpentier) (Odonata). *Entomologist* **79**: 171.

Longfield, C. 1947 *Coenagrion scitulum* (Rambur): a dragonfly new to the British list. *Entomologist* **80**: 54–5.

Longfield, C. 1949 The dragonflies of the London Area. *London Naturalist* for 1948. **28**: 80–98.

Longfield, C. 1954 The British dragonflies (Odonata) in 1952 and 1953. *Entomologist* **87**: 87–91.

Lucas, W.J. 1900a The dragonfly season of 1899. *Entomologist* **33**: 137–43.

Lucas, W.J. 1900b *British Dragonflies.* London: Upcott Gill.

Lucas, W.J. 1901 Odonata in 1900. *Entomologist* **34**: 65–9.

Lucas, W.J. 1902 Dragonflies at Navestock. *Entomologist* **35**: 116.

Lucas, W.J. 1927a Presumed immigration of the dragonfly *Sympetrum flaveolum*, Linn. in 1926. *Entomologist* **60**: 76–8.

Lucas, W.J. 1927b Notes on British Paraneuroptera in 1926. *Entomologist* **60**: 193–7.

McGeeney, A. 2004 Southern Hawker. In Brooks, S. (ed) *op. cit.*

McGeeney, A. 1997 Red-veined Darter *Sympetrum fonscolombei* (Selys) confirmed breeding in Britain in 1996, and notes on exuviae. *Journal of the British Dragonfly Society* **13** (2): 57–9.

McLachlan, R. 1871 Note appended to Doubleday 1871. *Entomologist's Monthly Magazine* **8**: 87.

McLachlan, R. 1884 The British dragon-flies annotated. *Entomologist's Monthly Magazine* **20**: 252–6.

McLachlan, R. & Eaton, A.E. 1870 *A Catalogue of the British Neuroptera.* Entomological Society of London.

Mays, R. 1978 *Henry Doubleday: The Epping Naturalist.* Precision Press.

Mearns, B. and Mearns, R. 2005 Four new species for South-west Scotland. *Dragonfly News.* **47**: 21.

Mendel, H. 1992 *Suffolk Dragonflies.*Ipswich: Suffolk Naturalist's Society.

Merritt, R., Moore, N.W. & Eversham, B.C. 1997 *Atlas of the Dragonflies of Britain and Ireland(ITE Research Publication 9).* Norwich: HMSO.

Miller, P.L. 1990 The rescue service provided by male *Enallagma cyathigerum* (Charpentier) for females after oviposition. *Journal of the British Dragonfly Society* **6** (1): 8–14.

Miller, P.L. 2004 Black-tailed Skimmer. In Brooks, S (ed) *op. cit.*

Morley, C. 1929 The dragon-flies of Suffolk. *Transactions of the Suffolk Naturalists' Society* **1**: 19–24.

Newman, E. 1832-3 Entomological notes. *Entomological Magazine* **1**–416 & 511–4.

Nobes, G. 2003 Southern Emerald Damselfly *Lestes barbarus* (Fabr.) – the first British record. *Atropos* **18**: 3–6.

O'Farrell, A.F. 1950 The J.J.F.X. King collection of British Odonata. *Entomologist* **83**: 14–8.

Parr, A.J. 1996 Dragonfly movement and migration in Britain and Ireland. *Journal of the British Dragonfly Society* **12** (2): 33–50.

Parr, A.J. 1997 Migrant and dispersive dragonflies in Britain during 1996. *Journal of the British Dragonfly Society* **13** (2): 41–8.

Parr, A.J. 1998 Migrant and dispersive dragonflies in Britain during 1997. *Journal of the British Dragonfly Society* **14** (2): 52–8.

Parr, A.J. 1999 Migrant and dispersive dragonflies in Britain during 1998. *Journal of the British Dragonfly Society* **15** (2): 51–8.

Parr, A.J. 2000 Migrant and dispersive dragonflies in Britain during 1999. *Journal of the British Dragonfly Society* **16** (2): 52–9.

Parr, A.J. 2001 Migrant and dispersive dragonflies in Britain during 2000. *Journal of the British Dragonfly Society* **17** (2): 49–54.

Parr, A.J. 2002 Migrant and dispersive dragonflies in Britain during 2001. *Journal of the British Dragonfly Society* **18** (1 & 2): 40–46.

Parr, A.J. 2003 Migrant and dispersive dragonflies in Britain during 2002. *Journal of the British Dragonfly Society* **19** (1 & 2): 8–15.

Parr, A.J. 2004 Migrant and dispersive dragonflies in Britain during 2003. *Journal of the British Dragonfly Society* **20** (2): 42–51.

Parr, A.J. 2005 Migrant and dispersive dragonflies in Britain during 2004. *Journal of the British Dragonfly Society* **21** (1): 14–21.

Parr, A.J. 2006 Migrant and dispersive dragonflies in Britain during 2005. *Journal of the British Dragonfly Society* **22** (1): 13–19.

Parr, A.J. 2004 First and last dates 2003/2004. *Dragonfly News* **46**: 19–20.

Parr, A.J. 2005a First dates for 2005. *Dragonfly News* **48**: 18.

Parr, A.J. 2005b Wildlife reports: dragonflies. *British Wildlife* **17** (2): 124–5.

Perrin, V.L. 1999 Observations on the distribution, ecology and behaviour of the Hairy Dragonfly *Brachytron pratense* (Müller). *Journal of the British Dragonfly Society* **15** (2): 39–45.

Perrin, V. 2005 Wildlife reports: dragonflies. *British Wildlife* **16** (5): 355–7.

Philpott, A.J. 1985 A large emergence of *Anax imperator*. *Journal of the British Dragonfly Society* **1**: 98–99.

Pinniger, E.B. 1933 Notes on the dragonflies of Epping Forest. *London Naturalist* for 1932: 66–72.

Pinniger, E.B. 1934a Epping Forest Odonata, 1933. *London Naturalist* for 1933: 98.

Pinniger, E.B. 1934b Notes on the habits of the dragonfly *Ischemia mixta* Latr. *London Naturalist* for 1933: 98.

Pinniger, E.B. 1935 Notes on dragonflies, 1934 *London Naturalist* for 193: 83.

Pinniger, E.B. 1936 Notes on dragonflies, 1935 *London Naturalist* for 1935: 71–2.

Pinniger, E.B. 1937 Notes on dragonflies. 1936. *London Naturalist* for 1936: 54.

Pinniger, E.B. 1938 Notes on dragonflies, 1937. *London Naturalist* for 1937: 77–9.

Pinniger, E.B. 1945 The Epping Forest Survey (3rd year) 6.1 Order Odonata. *London Naturalist* for 1944 **24**: 60–62.

Pinniger, E.B. 1947 *Coenagrion scitulum*, Rambur, a dragonfly new to Britain. *London Naturalist* for 1946 **26**: 80.

Pinniger, E.B. 1948 *Coenagrion scitulum*, Rambur (Odonata) in Essex. *Essex Naturalist* **28**: 69-71.

Pinniger, E.B., Syms, E. E. & Ward, B. T. 1950 Dragonflies in 1949. *Essex Naturalist* for 1949: **28**: 209–10.

Pittman, S. 1996 Migrant species of *Sympetrum* in Norfolk, 1995. *Journal of the British Dragonfly Society* **12** (1): 1–2.

Radford, A.P. 2000 Predation of a bumblebee (*Bombus* sp.) by the Four-spotted Chaser *Libellula quadrimaculata* L. *Journal of the British Dragonfly Society* **16** (2): 63.

Richardson, J.A. 1963 The Fingringhoe Wick ponds. *Nature in North East Essex*. 1962–3. CNHS.

Selys-Longchamps, E. de 1846a Revision of the British Libellulidae. *Annals and Magazine of Natural History*. Series 1. **18**: 217–27.

Selys-Longchamps, E. de 1846b Corrected list of British Dragonflies. *Zoologist* **4**: 1522–3.

Silsby, J. 1995 The 1995 darter invasion. *Newsletter of the British Dragonfly Society.* **28**: 11–13.

Silsby, J. & Ward-Smith, J. 1997 The influx of *Sympetrum flaveolum* (L.) during the summer of 1995. *Journal of the British Dradonfly Society* **13** (1): 14–22.

Stephens, J.F. 1835–7 *Illustrations of British entomology.* Vol 6. London: Baldwin and Cradock.

Swash, A. & Smallshire, D. 2004 *Britain's Dragonflies.*Hampshire: WildGuides.

Syms, E.E. 1929 Some biological notes on dragon-flies. *Essex Naturalist* **22**: 222–7.

Tarpey, T. & Heath, J. 1990 *Wild Flowers of North East Essex.* Colchester: CNHS.

Taylor, M.R. 1994 The predation of *Sympetrum sanguineum* (Müller) by *Vespula germanica* (Fabricius) (Hymenoptera, Vespidae). *Journal of the British Dragonfly Society* **10** (2): 39.

Thomas, A. 1999 *A Study of the Scarce Emerald Damselfly* (*Lestes dryas*) *and Other Species at Old Hall Marshes*, RSPB Nature Reserve. M.Sc. thesis, University College London.

Tillyard, R.J. 1917 *The Biology of Dragonflies.* Cambridge.

Tyrell, M. 2004 Group oviposition behaviour in the Brown Hawker *Aeshna grandis* (L.). *Journal of the British Dragonfly Society* **20** (2): 79.

Tyrrell, M. 2006 Observations on emergence and duration of adult life of the Hairy Dragonfly *Brachytron pratense* (Müller). *Journal of the British Dragonfly Society* **21** (2): 43–6.

Tyrrell, M. & Brayshaw, S. 2004 Population expansion of the Hairy Dragonfly *Brachytron pratense* (Müller) and other breeding dragonflies of the Nene Valley in Northamptonshire. *Journal of the British Dragonfly Society* **20** (2): 51–61.

Walker, F.A. 1897 Dragonflies of North London. *Entomologist* **30**: 120–22.

Ward, B.T. 1971 Personal view. *Essex Field Club Bulletin* Summer 1971: 5–8.

Wildermuth, H. 2000 Larvae of the Downy Emerald *Cordulia aenea* (L.) examine the space for eclosion with their hind legs. *Journal of the British Dragonfly Society* **16** (2): 59–62.

Winsland, D.C. 1995 Predation of emerging Odonata by the Black Ant (*Lasius niger* (L.)). *Journal of the British Dragonfly Society* **11** (2): 26–7.

Index

Note: This index includes only the page reference to the start of each species account, not every mention of the species in the book.

Aeshna cyanea 114
Aeshna grandis 118
Aeshna juncea 205
Aeshna mixta 121
Anax imperator 126
Anax parthenope 131
Azure Damselfly 89

Banded Demoiselle 65
Beautiful Demoiselle 60
Black Darter 207
Black-tailed Skimmer 150
Blue-tailed Damselfly 97
Brachytron pratense 109
Broad-bodied Chaser 137
Brown Hawker 118

Calopteryx splendens 60
Calopteryx virgo 65
Ceriagrion tenellum 199
Club-tailed Dragonfly 201
Coenagrion mercuriale 200
Coenagrion puella 89
Coenagrion pulchellum 204
Coenagrion scitulum 202
Common Blue Damselfly 93
Common Darter 162
Common Hawker 205
Cordulegaster boltonii 205
Cordulia aenea 133

Dainty Damselfly 202
Downy Emerald 133

Emerald Damselfly 69
Emperor Dragonfly 126
Enallagma cyathigerum 93
Erythromma najas 101
Erythromma viridulum 105

Four-spotted Chaser 146

Golden-ringed Dragonfly 205
Gomphus vulgatissimus 201

Hairy Dragonfly 109

Ischnura elegans 97
Ischnura pumilio 200

Keeled Skimmer 155

Large Red Damselfly 85
Lesser Emperor Dragonfly 131
Lestes barbarus 210
Lestes dryas 74
Lestes sponsa 69
Lestes virens 200
Libellula depressa 137
Libellula fulva 141
Libellula quadrimaculata 146

Leucorrhinia dubia 202

Migrant Hawker 121

Orange-spotted Emerald 201
Orthetrum cancellatum 150
Orthetrum coerulescens 155
Oxygastra curtisii 201

Platycnemis pennipes 81
Pyrrhosoma nymphula 85

Red-eyed Damselfly 101
Red-veined Darter 170
Ruddy Darter 157

Scarce Blue-tailed Damselfly 200
Scarce Chaser 141
Scarce Emerald Damselfly 74
Small Emerald Damselfly 200

Small Red Damselfly 199
Small Red-eyed Damselfly 105
Southern Damselfly 200
Southern Emerald 210
Southern Hawker 114
Sympetrum danae 207
Sympetrum flaveolum 166
Sympetrum fonscolombii 170
Sympetrum sanguineum 157
Sympetrum striolatum 162
Sympetrum vulgatum 206

Vagrant Darter 206
Variable Damselfly 204

White-faced Darter 202
White-legged Damselfly 81

Yellow-winged Darter 166